Applied Numerical Methods for Engineers

Applied Numerical Methods for Engineers

Contributors

Rekha R. Rao, Lisa A. Mondy et al.

AURIS
Reference

www.aurisreference.com

Applied Numerical Methods for Engineers

Contributors: Rekha R. Rao, Lisa A. Mondy et al.

Published by Auris Reference Limited

www.aurisreference.com

United Kingdom

Applied Numerical Methods for Engineers

ISBN: 978-1-78154-824-0

British Library Cataloguing in Publication Data

A CIP record for this book is available from the British Library

Printed in the United Kingdom
Exclusively distributed by CBS Publishers & Distributors Pvt. Ltd.
Sales & Distribution Rights only for India, Pakistan, Bangladesh, Sri Lanka, Nepal and Bhutan.This book is not to be sold outside these territories.

Contents

List of Abbreviations...vii

List of Contributors..ix

Preface.. xiii

Chapter 1   **3D Numerical Modelling of Mould Filling of a Coat Hanger Distributer and Rectangular Cavity**...1

Chapter 2   **Experimentally Validated Numerical Modeling of Heat Transfer in Granular Flow in Rotating Vessels**...31

Chapter 3   **Numerical Simulation of the Unsteady Shock Interaction of Blunt Body Flows**...83

Chapter 4   **Inverse Analysis Applied to Mushy Steel Rheological Properties Testing Using Hybrid Numerical-Analytical Model**..........................105

Chapter 5   **Optimization of Capacitive Acoustic Resonant Sensor Using Numerical Simulation and Design of Experiment**...........................139

Chapter 6   **Efficient Numerical Methods for Solving Differential Algebraic Equations**..167

Chapter 7   **Numerical Methods for Solving Turbulent Flows by Using Parallel Technologies**...179

Chapter 8   **A Review and Update of Analytical and Numerical Solutions of the Terzaghi One-Dimensional Consolidation Equation**.........................189

Chapter 9   **Numerical Study to Represent Non-Isothermal Melt-Crystallization Kinetics at Laser-Powder Cladding**...203

Chapter 10   **Convergence and Error of Some Numerical Methods for Solving a Convection-Diffusion Problem**...223

Chapter 11   **The Evolution of Pore Water Pressure in a Saturated Soil Layer between Two Draining Zones by Analytical and Numerical Methods**..243

Chapter 12   Design of Overall Slope Angle and Analysis of Rock Slope Stability
of Chadormalu Mine Using Empirical and Numerical Methods 263

Citations 273

Index... 275

List of Abbreviations

AST	Arvedi Steel Technology
CCD	Central composite design
DOE	Design of experiments
DAE	Differential algebraic Equation
DNS	Direct numerical simulation
DEM	Discrete Element Method
DRT	Ductility Recovery Temperature
FOS	Factor of Safety
FEM	Finite element method
GSI	Geological Strength Index
LES	Large eddy simulation
LC	Laser-powder cladding
NST	Nil strength temperature
PET	Polyethylene terephthalate
RSM	Response surface method
RMR	Rock Mass Rating
SQP	Sequential quadratic programming
SRMR	Slope Rock Mass Rating
SRT	Strength Recovery Temperature
TVD	Total-variation-diminishing

List of Contributors

Rekha R. Rao
Sandia National Laboratories USA

Lisa A. Mondy
Sandia National Laboratories USA

David R. Noble
Sandia National Laboratories USA

Matthew M. Hopkins
Sandia National Laboratories USA

Carlton F. Brooks
Sandia National Laboratories USA

Thomas A. Baer
Procter and Gamble Company USA

Bodhisattwa Chaudhuri
Department of Pharmaceutical Sciences, University of Connecticut, Storrs, CT, 06269

Fernando J. Muzzio
Department of Chemical and Biochemical Engineering, Rutgers University, Piscataway, NJ, 08854 United States of America

M. Silvina Tomassone
Department of Chemical and Biochemical Engineering, Rutgers University, Piscataway, NJ, 08854 United States of America

Leonid Bazyma
National Aerospace University "Kharkov Aviation Institute" Ukraine

Vasyl Rashkovan
National Polytechnic Institute Mexico

Vladimir Golovanevskiy
Western Australian School of Mines, Curtin University Australia

Miroslaw Glowacki
AGH University of Science and Technology Poland

Rubaiyet Iftekharul Haque
Centre Microélectronique de Provence, Ecole des Mines de Saint-Etienne, Gardanne 13541, France
TAGSYS RFID, 13600 La Ciotat, France

Christophe Loussert
TAGSYS RFID, 13600 La Ciotat, France

Michelle Sergent
Aix-Marseille Université, LISA EA 4672, 13397 Marseille Cedex 20, France

Patrick Benaben
Centre Microélectronique de Provence, Ecole des Mines de Saint-Etienne, Gardanne 13541, France

Xavier Boddaert
Centre Microélectronique de Provence, Ecole des Mines de Saint-Etienne, Gardanne 13541, France

Ampon Dhamacharoen
Department of Mathematics, Burapha University, Chonburi, Thailand

Alibek Issakhov
Department Mechanics and Mathematics, al-Farabi Kazakh National University, Almaty, Kazakhstan

Cheikhou Ndiaye
Laboratoire de Mécanique et Modélisation, UFR Sciences de l'Ingénieur, Université de Thiès, Thiès, Sénégal

Meissa Fall
Laboratoire de Mécanique et Modélisation, UFR Sciences de l'Ingénieur, Université de Thiès, Thiès, Sénégal

Mapathe Ndiaye
Laboratoire de Mécanique et Modélisation, UFR Sciences de l'Ingénieur, Université de Thiès, Thiès, Sénégal

Daouda Sangare
Laboratoire d'Analyse Numérique et d'Informatique, UFR Sciences Appliquées et Technologie, Université Gaston Berger, Saint-Louis, Sénégal

Abib Tall
Laboratoire de Mécanique et Modélisation, UFR Sciences de l'Ingénieur, Université

de Thiès, Thiès, Sénégal

V. G. Niziev
Institute on Laser and Information Technology Russian Academy of Sciences, Moscow, Russia

F. Kh. Mirzade
Institute on Laser and Information Technology Russian Academy of Sciences, Moscow, Russia

V. Ya. Panchenko
Institute on Laser and Information Technology Russian Academy of Sciences, Moscow, Russia

M. D. Khomenko
Institute on Laser and Information Technology Russian Academy of Sciences, Moscow, Russia

R. V. Grishaev
Institute on Laser and Information Technology Russian Academy of Sciences, Moscow, Russia

S. Pityana
CSIR-National Laser Centre, Pretoria, South Afric

C. V. Rooyen
CSIR-National Laser Centre, Pretoria, South Afric

Gabriela Nut
Applied Mathematics Department, Babes-Bolyai University, Cluj Napoca, Romania

Ioana Chiorean
Applied Mathematics Department, Babes-Bolyai University, Cluj Napoca, Romania

Petru Blaga
Applied Mathematics Department, Babes-Bolyai University, Cluj Napoca, Romania

Abib Tall
Laboratoire de Mécanique et Modélisation, UFR Sciences de l'Ingénieur, Université de Thiès, Thiès, Sénégal

Cheikh Mbow
Groupe de Recherches sur les dynamiques des Systèmes et la Mécanique des Fluides, Faculté des Sciences et Techniques, Université Cheikh Anta Diop, Dakar, Sénégal

Daouda Sangaré
Laboratoire d'Analyse Numérique et d'Informatique, UFR Sciences Appliquées et Technologie, Université

Mapathé Ndiaye
Laboratoire de Mécanique et Modélisation, UFR Sciences de l'Ingénieur, Université de Thiès, Thiès, Sénégal

Papa Sanou Faye
Laboratoire de Mécanique et Modélisation, UFR Sciences de l'Ingénieur, Université de Thiès, Thiès, Sénégal

Mahdi Rasouli Maleki
Engineering Geology & Rock Mechanic Department, Tunnel Consulting Engineers, Tehran, Iran

Mohammad Mahyar
Mining Engineering, Tunnel Consulting Engineers, Tehran, Iran

Kambiz Meshkabadi
Lecturer of Civil Engineering Department, Islamic Azad University of Ahar, Iran

Preface

Numerical methods are designed for the constructive solution of mathematical problems requiring particular numerical results, usually on a computer. A numerical method is a complete and unambiguous set of procedures for the solution of a problem, together with computable error estimates. The text Applied Numerical Methods for Engineers focuses on numerical methods for all engineering disciplines. First chapter focuses on 3D numerical modelling of mold filling of a coat hanger distributer and rectangular cavity. Experimentally validated numerical modeling of heat transfer in granular flow in rotating vessels has been presented in second chapter. Third chapter presents the results of numerical simulation of the flow around a hemisphere at both the symmetric and asymmetric energy supply into the flow, when the energy supply is realized at 900 angle to the velocity vector of the incoming supersonic airflow. The goal of fourth chapter is to present problems of theoretical work leading to the development of a methodology of very high temperature testing of steel samples while their central parts are still mushy. Fifth chapter outlines a new design scheme for acoustic sensor optimization that combines numerical simulation using the COMSOL multiphysics software and design of experiments (DOE) approach to optimize the acoustic sensor of the proposed design to obtain the acoustic resonator. Efficient numerical methods for solving differential algebraic equations have been discussed in sixth chapter. Seventh chapter focuses on numerical methods for solving turbulent flows by using parallel technologies. In eighth chapter, we resolve Terzaghi partial differential equations using Fourier method and finite difference respectively for analytical and numerical solutions. In ninth chapter, an improved phase change problem has been introduced to involve rapid melting and crystallization which is typical for laser-powder cladding (LC) technology. In tenth chapter, we present an eoretical study of the smoothing, convergence and error reduction properties of the multigrid method for a time dependent convection diffusion equation. The objective of eleventh chapter is to establish solutions, by analytical and numerical method, of the equation of the pore water pressure. The purpose of last chapter is to determine the bench slope angle and overall slope of the west wall in Chadormalu mine in points susceptible to rupture.

Chapter 1

3D NUMERICAL MODELLING OF MOULD FILLING OF A COAT HANGER DISTRIBUTER AND RECTANGULAR CAVITY

Rekha R. Rao[1], Lisa A. Mondy[1], David R. Noble[1], Matthew M. Hopkins[1], Carlton F. Brooks[1] and Thomas A. Baer[2]

[1]Sandia National Laboratories USA

[2]Procter and Gamble Company USA

INTRODUCTION

Filling processes occur in a wide range of industries, ranging from packaging of consumer products to manufacturing processes for making polymeric, metal and ceramic components. These processes involve the complex interplay of extrusion of a viscous liquid into a mould or container where it displaces a gas phase. Numerical modeling based on computational fluid dynamics can be useful for understanding the filling process. However, complexity arises in that the fluid dynamics in both the viscous liquid and gas phase must be resolved while concurrently determining the location of the fluid-gas interface and the interaction of this interface with the solid surface, i.e., the wetting behavior. Determining the free surface location and wetting behavior is an integral part of the numerical method.

Numerical methods have been applied to bottle and container filling for consumer products where the rheology can include shear thinning and viscoelastic effects and instabilities such as buckling and coiling may be prevalent [Tome, et al., 2001; Oishi et al., 2008; Roberts & Rao, 2011; Ville et al., 2011]. In metal casting simulations, the fluids are generally Newtonian, but complexity arises from the high injection rates leading to turbulent flow and temperature-dependent behavior such as solidification [Ilinca & Hetu, 2000; Cross et al, 2006]. For injection molding of polymers, time- and temperature-dependent effects are seen in conjunction with non-Newtonian rheology [Ilinca & Hetu, 2001; Kumar & Ghoshdastidar, 2002]. In powder injection molding for ceramic and metal forming, a suspension of particles is injected into a mould to create a green part, which later sees further processing steps to produce the final part [Hwang & Kwon, 2002; Ilinca & Hetu, 2008]. Numerical methods for these problems range from finite difference, to finite volume, and finite element.

General classes of algorithms for determining the location of the free surface include Eulerian, Lagrangian and arbitrary Lagrangian-Eulerian (ALE) descriptions. Eulerian methods use a fixed-grid with an interface capturing technique such as the volume of fluid [Hirt & Nichols, 1981] or level-set method [Sethian, 1999] to determine the location of the free surface. Traditional Lagrangian methods use a moving mesh as a material interface that advects with the fluid.

These methods often require multiple remeshing steps to avoid mesh distortion and tangling [see for instance, Bach & Hassager, 1985; R. Radovitzky & M. Ortiz, 1998; Zhang & Khayat, 2001]. To avoid these problems, Lagrangian mesh-free methods have been developed using smooth particle hydrodynamics or the material point method to avoid meshing issues [Kulasegaram et al, 2006; Kauzlaric et a;., 2011; Love & Sulsky, 2006]. These methods work well for being able to capture a moving interface with topological changes, but have difficulty in accurately solving the base physics, e.g. viscosity, and applying boundary conditions such as surface tension. ALE methods are hybrid techniques that seek to exploit the benefits of both the Eulerian and Lagrangian description in a hybrid manner, to determine the location of the interface [i.e. Sackinger et al, 1996; Lewis et al., 1998; Nithiarasu, 2005].

Injection loading of a ceramic paste is a high-rate process used to create green ceramic parts, which subsequently experience binder burnout and sintering to produce the final ceramic part. In the process of interest, the mould is a rectangular cavity, with an inflow from a coat hanger die that should distribute the flow evenly across the mould inflow. The cavity is small with dimensions of 1.3 cm by 3.6 cm in plane and a height of 0.4 cm. The inflow tube to the distributer has a diameter of 0.5cm. Figure 1 shows short shots (or incomplete filling of a mould) for the injection loading process, illustrating some of the defects that can occur when a fluid with complex shear and temperature-dependent rheology interacts with a high rate process.

Temperature control and high-rate processing can limit the folding instability seen on the left of Figure 1. However, the pooling phenomenon shown on the right of Figure 1 occurs when optimal processing conditions are used, indicating that the design of the distributer may be responsible for material building up in the centre of the die.

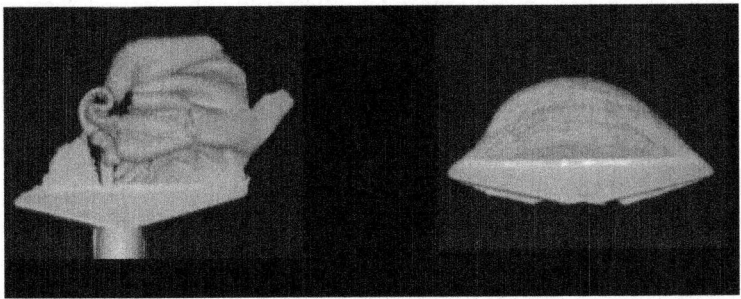

Figure 1: Short shots of injection loading of a ceramic paste in a part are shown [Rao et al, 2006]. The photo on the left is injection loading using a slow injection speed, while the photo on the right uses a higher injection speed. At low filling speed, the paste acts like a solid material. Even at high filling rates, when the paste begins to act as a fluid, pooling at the center of the mould is seen and the desired mould shape is not achieved.

The complex shear-rate and temperature-dependent rheology of the ceramic paste was determined to follow a power-law dependence on shear rate and a Williams-Landau-Ferry temperature dependence [Rao et al., 2006]. The material shear thinned quickly to a constant viscosity value at moderate shear rates. Thus, because of the high shear rates in the injection loader, it was determined that the rheology was essentially Newtonian as long as the temperature remained constant at the processing temperature. Therefore, a study was undertaken to better understand the filling dynamics and reduce pooling by changing the distributer design using a Newtonian fluid.

DYNAMIC WETTING MODELS AND MOULD FILLING

Understanding dynamic wetting, or the interaction between the free surface and the mould walls, has been the subject of numerous experimental and theoretical studies and is still an outstanding research topic [see for instance Blake, 2006; Ren et al, 2011]. The difficulty arises from the contradiction of a moving contact line at the fluid-gas interface and the no slip boundary conditions traditionally applied at solid surfaces. How can the contact line advance when the velocity vanishes at the solid surface? Highly viscous materials such as polymers and particle suspensions have large capillary numbers (a measure of the ratio of viscous forces to capillary forces) and are often hypothesized to obey a rolling motion condition with an 180o dynamic contact angle. Numerically, this approach is difficult to apply and other methods have been proposed. The simplest models use a Navier slip condition to allow either slip on the entire solid surface, slip for the gas phase only, or slip only at the dynamic contact line.

These models are ad hoc and ignore any thermodynamics considerations such as the static contact angle and surface energy. More advanced wetting models generally give dynamic contact angle as a function of the local Ca, the static contact angle and other material properties of the fluid and the solid surface [see Schunk et al., 2006 for a brief review of this work]. Hoffman [1975] used experimental measurements to develop a universal correlation for the dynamic contact angle as a function of static contact angle and a local capillary number, while Cox [1986] developed a competing model using asymptotic analysis. Kistler [1983] used a linear model that was easy to implement in numerical computer codes. Shikhmurzaev [1994] used hydrodynamic theory and included a surface phase as part of the wetting model. Blake [1969] developed a molecular kinetic theory that reduces to a linear model for small contact angles. Blake noted that the advancing angle is a monotonically increasing function of Ca. The degree of velocity dependence will however increase steeply as viscosity increases or surface tension decreases.

To model dynamic contact for the filling process, the dynamic angle is tied to the balance of forces at the advancing wetting line, namely the tangential wetting line force, liquid-gas surface tension force, and fluid viscous force. Here, a version of the Blake model is used that is straightforward to populate with experiments. The dynamic contact angle is measured in the laboratory as a function of the velocity of the wetting line for the fluids and surface used in our experiments as input to the wetting model. The Blake model is also easy to implement numerically, since the wetting speed can be written as a function of the dynamic and static contact angles. The performance of the Blake model at the high capillary number limit may be suspect, since it exhibits an unbounded dynamic contact angle while more physical models such as Cox and Hoffman reach a limit of 180o for large capillary numbers [Schunk et al, 2006].

MOULD DESIGN

Because the initial design of the mould exhibited pooling of the fluid in the centre of the cavity and this would lead to poor filling (see Figure 1), we tried two minor redesigns to the distributor to see if we could improve the flow into the mould and reduce pooling. Ideally, we would like to see a flat profile coming out of the distributor and a more one-dimensional front shape. Figure 2 shows the original mesh, Mesh 1, and a variation, Mesh 2, with a longer distributor, and Mesh 3 with a longer-taller distributor. The idea behind Mesh 2 was to give a longer length for the flow to develop a flat profile and fill up the distributor before entering the main cavity. Mesh 3 kept this longer distributor and made it wider on inflow to ease the fluid entering the cavity. These ideas were inspired by discussions in Sartor [1990] about die design. The meshes themselves all have the same cavity size and a similar amount of refinement, though Mesh 2 and Mesh 3 have more elements and unknowns due to the longer distributor.

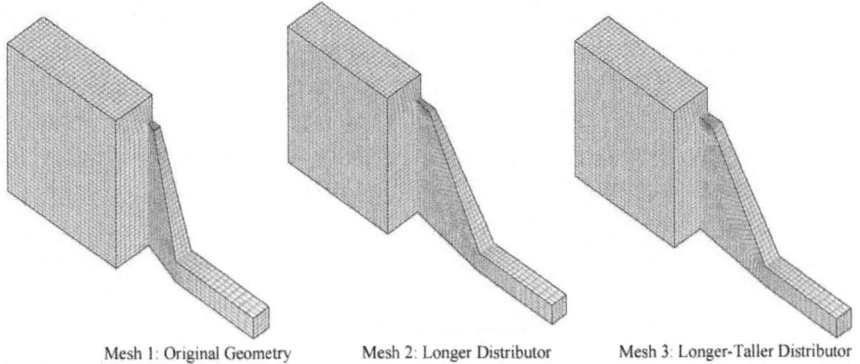

Mesh 1: Original Geometry   Mesh 2: Longer Distributor   Mesh 3: Longer-Taller Distributor

Figure 2: Geometries to be investigated: Mesh 1 is the original design, Mesh 2 incorporates a longer distributer, and Mesh 3 incorporates a longer distributer with a wider inflow to the cavity.

CHAPTER ORGANIZATION

In this chapter, we investigate the design of an injection loading mould using flow visualization experiments and numerical models. The finite element method is used to understand the interaction of the inflowing viscous liquid with the geometry and the displaced gas phase. Filling dynamics are determined with a diffuse-interface implementation of the level set method [Sethian 1999]. The Blake wetting model is used to represent the interaction of the free surface with the mould surface at the dynamic contact line [Blake & Haynes, 1969; Blake, 2006]. The flow visualization experiments are carried out under isothermal conditions using an acrylic mould and a viscous Newtonian fluid. Three different moulds are examined in two different orientations with gravity. Simulation results are given for these six cases.

The chapter is organized in the following manner. First, the equations and numerical method are presented. Next, the experimental methods used to provide input parameters for the models and flow visualization studies to better understand the filling dynamics and provide confidence in the numerical method are discussed. In the subsequent section, the results for injection molding process are given for a Newtonian fluid into a coat hanger die distributer and a rectangular cavity, where the 3D level set simulations are compared to experiments. We conclude by summarizing the results and discussing future efforts.

EQUATIONS AND METHOD

Equations of Motion

We can write the equations of motion for a single-phase fluid and then generalize them for our multiphase flow problem, where a viscous fluid displaces a gas. The fluids of interest are assumed to be incompressible and have a constant density, meaning that the velocity field, u, will be solenoidal and the continuity equation contains no density or pressure variables. (Note, this is a good assumption for the viscous liquid but a simplification for the gas phase, which is actually compressible.)

$$\nabla \cdot u = 0$$

(1)

Conservation of momentum for a Newtonian fluid takes into account gradients in the fluid stress tensor, T, defined as the product of the viscosity m and the shear rate tensor, ( )t Ñ +Ñ u u , gradients in the pressure, p, as well as gravitational effects and inertial terms that can be dependent on time, t, and the fluid density, r. Note that gravity, g, can be an important body force in filling processes and most filling processes fill counter to gravity.

$$\rho(\frac{\partial u}{\partial t} + u \cdot \nabla u) = \mu \nabla^2 u - \nabla p + \rho g$$

(2)

Because we have a viscous fluid displacing a gas phase, the location of the interface between fluids is unknown a priori. To determine the location of the free surface as it evolves in time, we use an Eulerian interface-capturing scheme based on the level set method, the details of which are included in the following section.

Interface Capturing

We use the level set method of Sethian [1999] to determine the evolution of the interface with time. The level set is a scalar distance function, the zero of which coincides with the free surface or fluid-gas interface, e.g.

$$\phi(x, y, z) = 0$$

(3)

We initialize this function to have a zero value at the fluid-gas interface, with negative distances residing in the fluid phase and positive distances in the gas phase. An advection equation is then used to determine the location of the interface over time.

$$\frac{\partial \phi}{\partial t} + v \cdot \nabla \phi = 0$$

$$(4)$$

Derivatives of the level set function can give us surface normal, n, and curvature, k, at the interface, which is useful for applying boundary conditions.

$$n = \frac{\nabla \phi}{|\nabla \phi|}$$

$$\kappa = \frac{-\nabla^2 \phi}{|\nabla \phi|}$$

$$(5)$$

We use the equations of motion described in the previous section, but vary the material properties across the phase interface. This variation is handled using a smooth Heaviside function that modulates material properties to account for the change in phase.

$$H_{gas}(\phi) = \frac{1}{2}(1 + \frac{\phi}{\alpha} + \frac{1}{\pi}\sin(\frac{\pi\phi}{\alpha})); -\alpha \le \phi \le \alpha$$

$$H_{fluid}(\phi) = 1 - H_{gas}(\phi)$$

$$(6)$$

This is a diffuse interface implementation of the level set method, which allows for an interfacial zone of length 2a. This zone is usually chosen to be four to six elements wide. Equation averaging is done using a Heaviside for the gas and viscous fluid equations. Because the properties are linear, this process results in Heaviside-averaged properties in a single momentum equation and an unchanged continuity equation:

$$\rho_{average}(\frac{\partial u}{\partial t} + u \cdot \nabla u) = \mu_{average}\nabla^2 u - \nabla p + \rho_{average}g$$

$$\nabla \cdot u = 0$$

$$(7)$$

The properties have fluid properties in some regions and gas properties in other regions as modulated by the numerical Heaviside. In the diffuse interface region, the properties are averaged between fluid and gas values. Figure 3 shows a schematic representation of the Heaviside function.

$$\rho_{average} = H_{gas}\rho_{gas} + H_{fluid}\rho_{fluid}$$

$$\mu_{average} = H_{gas}\mu_{gas} + H_{fluid}\mu_{fluid}$$

(8)

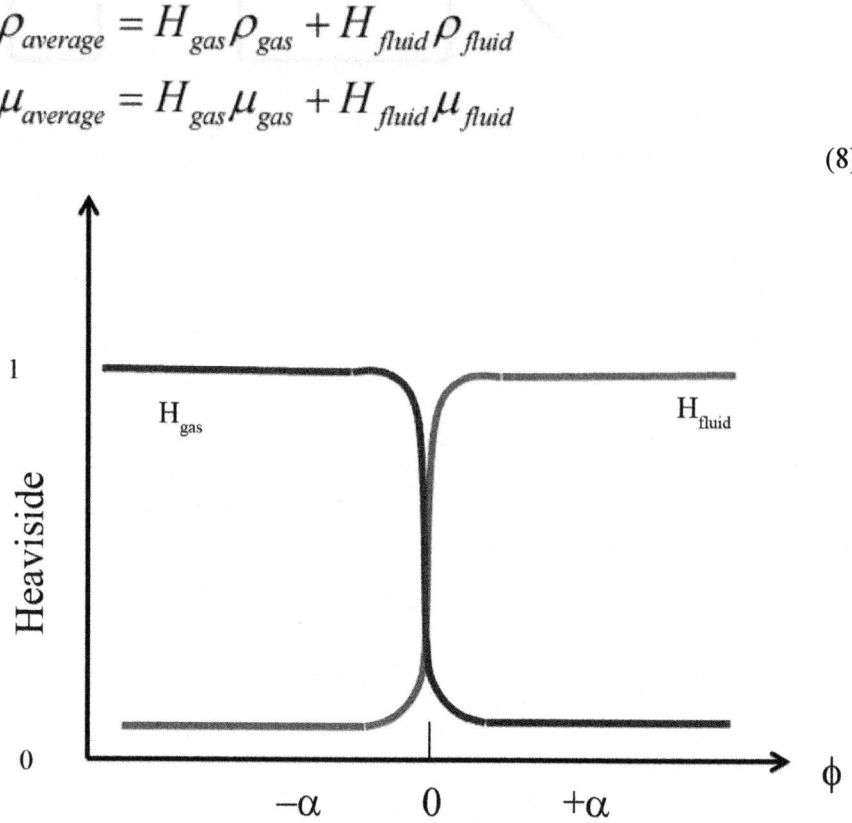

Figure 3: Numerical Heaviside for averaging material properties.

The regularized Dirac delta function, which is defined as

$$\delta_\alpha(\phi) = \frac{dH(\phi)}{d\phi} = \frac{|\phi|}{2\alpha}(1+\cos(\frac{\pi\phi}{\alpha})),$$

(9)

is used to apply surface tension and capillary boundary conditions via a continuous surface force approach [Brackbill et al, 1992]. This applies surface tension as a volumetric body force on the momentum equation, which is distributed throughout the interfacial zone region through the regularized Dirac delta function.

$$(\mu_{gas} - \mu_{fluid})n\bullet(\nabla u + (\nabla u)') = 2\sigma\delta_\alpha(\phi)\kappa n$$

(10)

Finite element implementation

The equations of motion (7) and the level set advection (4) were solved using a finite element method as implemented in ARIA [Notz et al., 2006]. Bilinear shape functions were used for the three velocity components, pressure, and level set. The LBB requirement on the velocity and pressure space was circumvented using Dohrmann-Bochev pressure stabilized pressure-projection (PSPP) [Dohrmann and Bochev, 2004] to allow for this equal order, bilinear, interpolation of all variables. (LBB compliant elements have the velocity space higher than the pressure space [Hughes, 2000].) The velocity vector and pressure unknowns were solved in the same matrix, while the level set equation was solved in a separate matrix, but at the same time step intervals. The level set equation was stabilized using a Taylor-Galerkin method [Donea, 1985]. The PSPP stabilization method greatly improved the condition number of the discretized matrix equations when compared to LBB elements or other stabilization methods, allowing for the use of an ILU preconditioner with a BiCGStab Krylov iterative solver. Further details of the modelling approach and equations, the numerical methods used and the finite element implementation can be found in Rao et al. [2011].

Contact-line Wetting Model

Boundary conditions for the dynamic contact line where the free surface and wall intersects are handled with a Blake wetting condition [Blake and Haynes, 1969; Blake, 2006]. Parameters for the model are informed by goniometer experiments that determine the wetting speed, v_{wet} as a function of the dynamic contact angle, θ, and the static contact angle, θ_s for the various fluids and surfaces of interest [Mondy et al., 2007].

$$v_{wet} = v_o \sinh(\overline{\gamma}(\cos\theta_s - \cos\theta)) - \tau\frac{\partial v_{wet}}{\partial t}$$

(11)

The dynamic contact angle can be calculated from the level set function.

$$\cos\theta = n_w \cdot \nabla\phi / |\nabla\phi|$$

(12)

When we integrate the stress, T, by parts in the finite element implementation of the momentum equation, a surface term is created as a natural boundary condition. We exploit the surface stress term and add on a Navier slip condition that includes the wetting speed from the Blake model.

$$\vec{n} \cdot T = -\frac{1}{\beta}\left(-\tau\frac{\partial \vec{v}}{\partial t} + \vec{v} - f_\alpha(\phi)v_{wet}\vec{t}\right)$$

$$f_\alpha(\phi) = 1 - \frac{|\phi|}{\alpha}, -\alpha \le \phi < \alpha; 0, |\phi| > \alpha$$

(13)

The value of the tangential wall velocity ramps from zero at a level set length scale away from the contact line to v_{wet} from equation (11) at the contact line. Away from the contact line, we revert to no slip for the tangential wall velocity boundary condition. The normal velocity is enforced as no penetration everywhere. The shape of this ramp must be smooth in order to get a realistic wetting line that shows a smooth transition from the contact line to the bulk flow. For a sharp transition from slip to Blake, we ended up with unphysical looking cusps in the interface shape near the wall. The transient terms introduce dynamics into the wetting line motion and allow smooth movement of the contact line. The t parameter is taken to be approximately the time step. The b parameter is generally small so that equation (13) functions almost like a Dirichlet requirement on the fluid velocity.

EXPERIMENTS

It was decided to visually record the flow of a simple, single phase, Newtonian liquid through transparent moulds to build confidence in the front tracking and wetting models used in the computations. Details of the geometries, experimental conditions, and properties of the materials used in each test will be given in the following sections.

Geometries

For these tests, we built transparent acrylic moulds identical to the three mesh geometries (Figure 2). The strength of the acrylic material making up the transparent moulds dictated that we inject at a lower injection pressure and lower operating temperature than the actual injection loading process. To mimic the actual injection process, we used a pressure driven syringe held at a constant pressure of 29.95 ± 0.10 psig during injection. The syringe of liquid was degassed in a vacuum chamber prior to the experiment. The syringe was modified to prevent leakage around the plunger and subsequent bubble formation. However, this resulted in more friction and a flow rate that varied somewhat from experiment to experiment. Hence, the time to fill the moulds varied from test to test, but was determined and recorded for each test. Reported below is the median and spread of the measured filling time for four repeated experiments in each geometry.

The tests were conducted at a room temperature that ranged from 23 to 24°C. To minimize the effect of temperature on the viscosity, the liquid was held in a water bath set to 23.5°C after degassing and before being used in an experiment. These conditions resulted in an injection rate that was approximately ten times slower than the actual process. However, the Reynolds number for the actual process and the validation experiments were similar and in the Stokes regime of much smaller than one. For an average fill time of 20 s, a cavity volume of 1.87 cm³, and characteristic length scale from the inflow port of 0.5cm, the Reynolds number was 0.0006 and the capillary number was 4.5.

Materials and Properties

We chose a liquid, UCON 75-H-90,000 (oxyethylene/oxypropylenes from Dow Chemical) as our model fluid. This UCON was chosen since its viscosity is an order of magnitude lower than the ceramic paste at processing conditions. Combined with an order of magnitude decrease in the injection rate compared to the real system, this gives a similar Reynolds number. The liquid surface tension and wetting properties of the UCON lubricant were determined at 23.5°C. The viscosity of the lubricant was measured with a Rheologica™ constant stress rheometer and was equal to 390 Poise over a shear rate ranging from 0.1 to 10 sec-1, indicating Newtonian rheological behaviour. The density of the liquid was measured with a densitometer to be 1.09 g/cm³. The surface tension measured with a Du Noüy ring (mean circumference of 5.935 cm) was 42.4±0.1 dyne/cm.

The dynamic contact angle on acrylic was measured with a feed-through goniometer [Mondy et al., 2007], in which liquid can be continuously injected to achieve "high" velocities, or the sessile drop can be allowed to relax to obtain "low" velocities. Figure 4 shows a schematic of the experimental apparatus and the results of this wetting test. During each experiment, the angle of contact and the location of the triple point of contact were recorded. The wetting line speed was then determined from the location of the triple point with time. The data fit to the Blake model (equation 9) gives a wetting constant vo of 0.00130cm/s and scale factor g of 2.29 with a static contact angle θ of 37.3°.

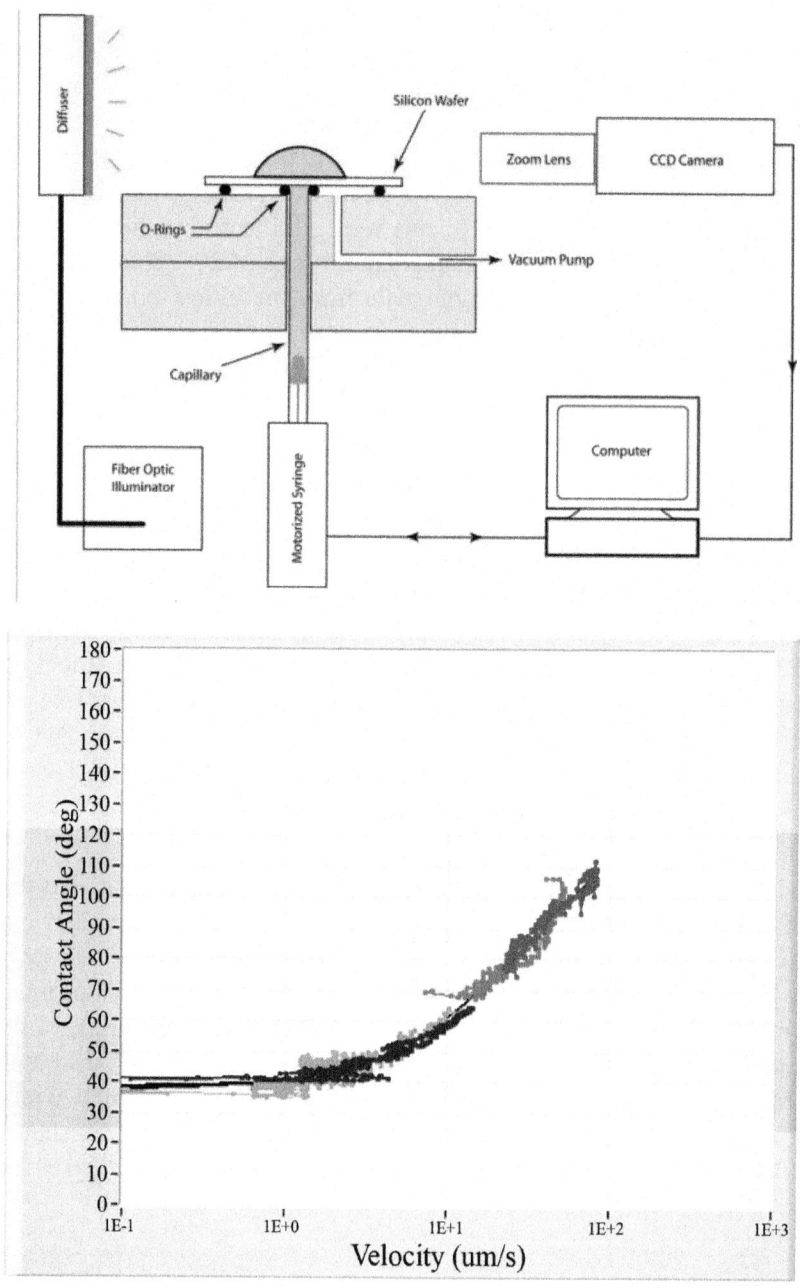

Figure 4: Sketch of apparatus for wetting parameter measurements (left). Dynamic wetting measurements of contact angle vs. velocity for 75-H-90000 UCON on acrylic (right).

Flow tests

Figures 5 through 7 are representative frames from the video recordings of the fill process using a vertical alignment of the mould as in the proposed ceramic injection process. The time it took to completely fill the original geometry, Mesh 1, (defined by the time when the liquid began to exit along the entire length of the vent) was 24.6 ± 1.2 s, to fill the geometry of Mesh 2 was 26.9 ± 2.0 s, and to fill the geometry of Mesh 3 was 24.6 ± 2.3 s. Because the fill rates varied, the time is shown in these figures in a nondimensional form. The initial time is taken to be when the front passes a line in the square entry channel 0.16 cm from the entrance of the distributor. Here, one can see the effects of changing the distributor geometry. In Figure 5 the liquid has just filled the distributor. All of the modified geometries help flatten the leading front, especially Mesh 2. Mesh 2 fills the distributor with the fastest relative time

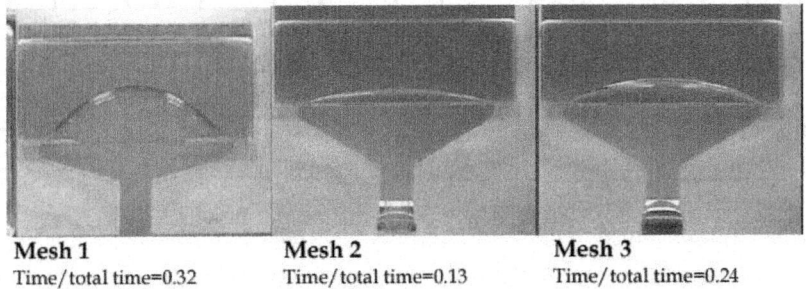

Mesh 1   **Mesh 2**   **Mesh 3**
Time/total time=0.32   Time/total time=0.13   Time/total time=0.24

Figure 5: Comparison of the effect of distributor geometry on the shape of the fluid front entering the mould for vertical mould orientation.

Figure 6 shows the times at which the leading front hits the wall opposite the injection port. In this case, the front in Mesh 2 takes the longest relative time to reach this stage. The other geometries once again follow the same pattern as before, with Mesh 2 giving the flattest flow profile and the original mesh the displaying the most curvature of the interface. At this point, the fluid has wetted more of the side walls with the Mesh 2 geometry and the front is flatter. Because Mesh 2 has a flatter profile, it takes the longest time to reach the top wall compared to Mesh 1 and Mesh 3.

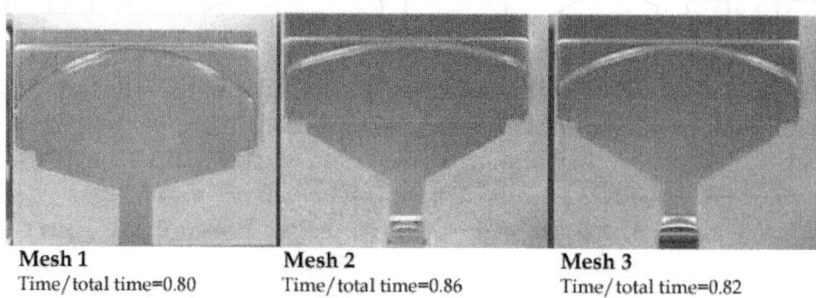

| Mesh 1 | Mesh 2 | Mesh 3 |
| Time/total time=0.80 | Time/total time=0.86 | Time/total time=0.82 |

Figure 6: Comparison of the effect of distributor geometry on the time it takes to reach the wall farthest from the injection port for vertical mould orientation.

Figure 7 shows the locations of the voids remaining in the mould once filled but without any over pressure (a relative time of 1). Small bubbles away from the corners are artefacts of the syringe loading process. The voids in geometries with redesigned distributors remain in the same locations as those seen in the original geometry. The relative areas of the bubbles on the images, which reflect the volume of air trapped, were determined and compared quantitatively in Table 1. One pixel resolution represents about 1×10^{-5} cm^2. All redesigned distributors result in smaller bubbles in the upper corners than those in the original mesh. The lower bubbles of Mesh 2 are also smaller, whereas those in Mesh 3 are approximately the same as those in the original geometry.

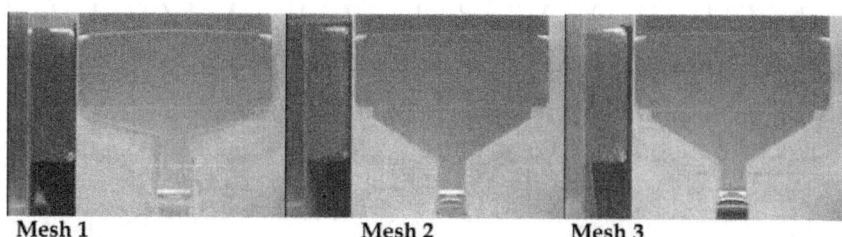

Mesh 1   Mesh 2   Mesh 3

Figure 7: Voids remain in the front upper and lower corners of each geometry for the vertical mould orientation. Each frame consists of two views of the mould (top view and side view). These voids can be seen more easily from the side view.

Table 1: Bubble sizes remaining in the corners for vertical mould orientation.

Geometry	Area Upper Corners (cm²)	Area Lower Corners (cm²)
Original	0.0153 ± 0.0012	0.0020 ± 0.0005
Mesh 2	0.0086 ± 0.0007	0.0012 ± 0.0002
Mesh 3	0.0128 ± 0.0005	0.0021 ± 0.0001

Next, we experimented with the orientation of the mould. Moulds were turned so that the distributor was perpendicular to gravity on the lower surface and the vent was moved to be on the upper surface. In other words, the flow direction and gravity are now perpendicular, with gravity acting in the thinnest cavity direction. Results for the original mesh showed that orientation with respect to gravity had a large impact on the likelihood of voids remaining in the corners of the mould. Figure 8 shows representative video frames at three stages of the fill: 1) when the distributor was completely filled, 2) when the front hit the far wall, and 3) when the liquid began to exit through the vent and the sides had completely wetted. Because the fill rates varied, the time is shown in a nondimensional form. The time it took to completely fill the original geometry in the horizontal position was 23.3 ± 0.9 s, to fill the geometry of Mesh 2 was 28.9 ± 1.5 s, and to fill the geometry of Mesh 3 was 27.3 ± 1.8 s.

The top views of the horizontal orientation show that the liquid wets the top later than the bottom. The oval area of the middle image in Figure 8 indicates where the liquid has wetted the upper surface. The side views are more difficult to interpret because of the lighting challenges accompanying trying to see through the thickest section of the mould. However, the dark areas of the side view are where the liquid has wetted the sides. It is interesting to note that the horizontal alignment causes the front to hit the far wall before even wetting the sides at all. In other words, the front was less "flat" entering the mould from the distributor. Nevertheless, the results also show that the horizontal orientation, with the vent on top and the distributor entrance on bottom, resulted in bubbles only in the upper corners nearest the distributor. When oriented vertically, bubbles are trapped in corners both opposite the distributor and opposite the vent. The bubbles nearest the distributor with the horizontal orientation are approximately the same size as the bubbles trapped in the upper corners opposite the vent in the vertical orientation. The effect of distributor geometry on the filling in the horizontal orientation is shown in Figure 9 and Figure 10, which correspond to Figure 5 and Figure 6 in the vertical orientation. Again, the geometry of Mesh 2 results in the flattest front entering the mould from the distributor (Figure 9). The front reaches the back wall at approximately the same time using the distributors of Mesh 2 and 3. By the time the front reaches the back wall both modified geometries result in more liquid filling the mould than with the original geometry; however, Mesh 2 results in somewhat more than Mesh 3 (Figure 10).

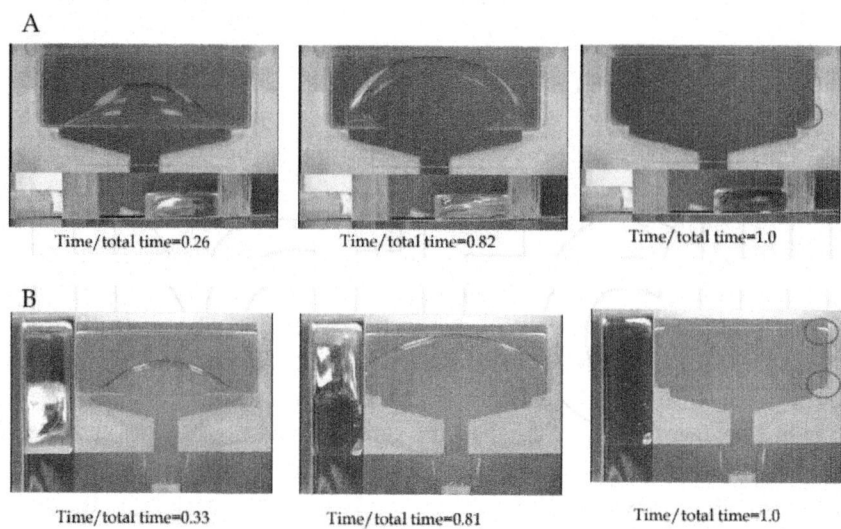

Figure 8: Original geometry (Mesh 1) oriented horizontally (A) and vertically (B). Images on the left show when the distributor fills completely, middle images show when the front first hits the back wall, and images on the right show when the part is filled to the point that the fluid leaves the vent area through the entire length of the vent. Bubbles left in the corners are circled in red.

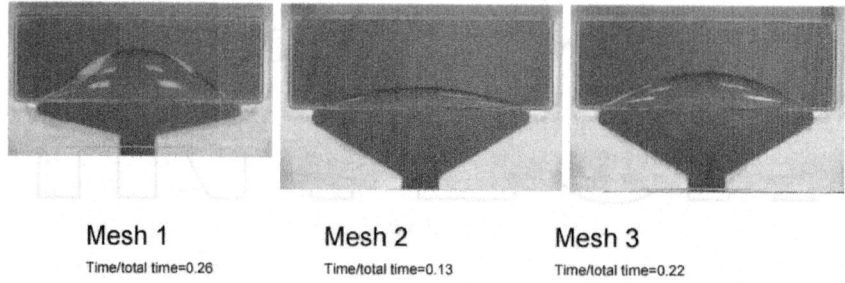

Figure 9: Comparison of the effect of distributor geometry on the shape of the fluid front entering the mould in a horizontal orientation. The flow direction and gravity are perpendicular, with gravity acting in the thinnest cavity direction.

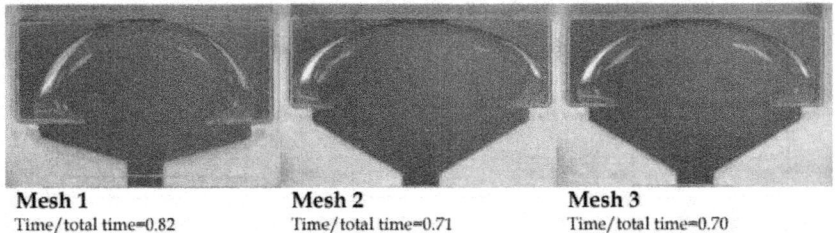

Mesh 1   Mesh 2   Mesh 3
Time/total time=0.82   Time/total time=0.71   Time/total time=0.70

Figure 10: Comparison of the effect of distributor geometry on the time it takes to reach the wall farthest from the injection port in a horizontal orientation.

Sizes of the bubbles left in the various geometries in the horizontal orientation are listed in Table 2. Comparing these values with the measurements in Table 1, one can see that the amount of gas left in the vertical orientation is roughly the same as that in the horizontal orientation, although in the horizontal orientation there are fewer bubbles. Bubbles observed in other locations than the corners are almost always artefacts of the syringe loading process.

Table 2: Bubble sizes remaining in horizontal orientation

Geometry	Area (cm^2)
Original	0.0189 ± 0.0002
Mesh 2	0.0111 ± 0.0002
Mesh 3	0.0123 ± 0.0017

Simulations The simulations runs were made on 64 processors of Thunderbird (Sandia National Laboratories capacity computing platform) and ran in less than 3.5 hours. This allowed us to do real time design and sensitivity calculations for parameters such as wetting speed, second phase viscosity, level set length scale and inflow pressure. Properties for the liquid used for the validation simulations were the measured values discussed in the experimental section above. The density and viscosity of the displaced gas phase were taken as fictitious values of one thousand times smaller than the liquid phase density and viscosity. These values are summarized in Table 3.

Table 3: Material properties used for validation simulations

Material Property	Value
Density of liquid	1.09 g/cm^3
Viscosity of liquid	390 Poise
Density of gas	0.0011 g/cm^3
Viscosity of gas	0.39 Poise
Wetting speed, v_o	0.0013 cm/s
Blake scale factor, γ	2.29
Static contact angle	37.3^o
Surface tension	42.4 dyne/cm
Inflow pressure	1.0x10^6 dyne/cm^2

At 25°C, the viscosity of air is 2.0x10^{-4} Poise and the density of air is 0.0012 g/cm^3, thus our second phase properties are very close for density, but three orders of magnitude too high for viscosity. The numerical method fails to converge for values of the liquid/gas viscosity ratio of more than 1000 for a diffuse interface implementation of the level set equations, so this is a necessary expedient requiring us to adhere to this fictitious viscosity value.

For the inflow condition, we used a constant inflow pressure of 1.0x10^6 dyne/cm^2. This value was chosen to match the horizontal fill time of 23 seconds in the original mesh and then used for all other meshes and geometries. A shooting method was used, where different values of the inflow pressure were used and the solution was examined to see if it filled in the correct time. This required many simulations to be run. It is believed that the actual boundary condition for the experiments is somewhere between a constant velocity and constant pressure condition, but this is hard to replicate numerically. From the simulations, we found that the velocity changed quickly in the beginning for the pressure inflow boundary condition and subsequently reached a steady value.

The initial 3D mesh and boundary conditions are given Figure 11 for the mould filling simulations. We assume symmetry about the centreline and only solve half the problem to improve the computational efficiency. The mesh contains 6744 8-Node hexahedral elements giving 41300 total degrees of freedom for bilinear velocity/bilinear pressure interpolation. This mesh was shown to be adequate, as a more refined version of this mesh gave the same fill times and meniscus shapes [Rao et al., 2006].

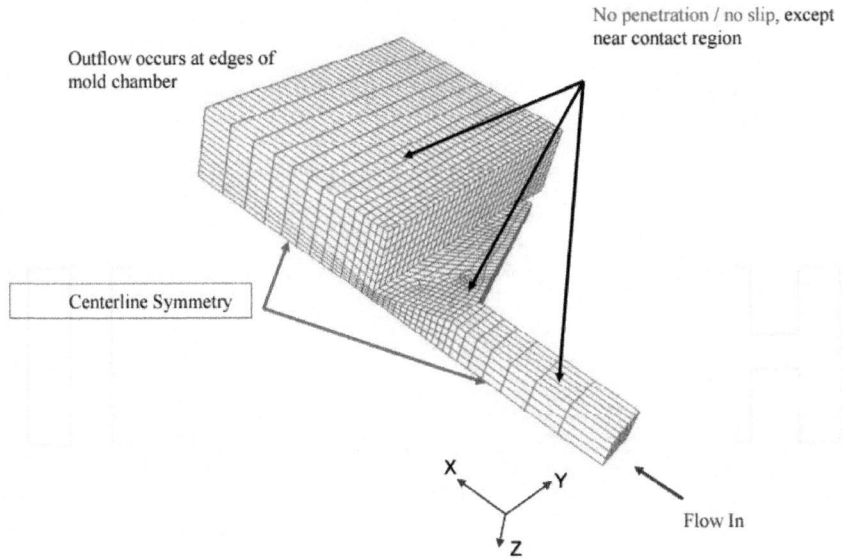

Figure 11: Initial mesh and boundary conditions for 3D level set simulations. The outflow vent is on the same side as the distributor for vertical simulations and opposite the distributor for horizontal simulations.

The time to fill the mould for the vertical orientation for Mesh 1, Mesh 2, and Mesh 3 are 15.2s, 17.5s, and 13.2s, which was much faster than the experimental values of 24.6s, 26.9s, and 24.6s. We defined our fill time as when the vent had filled completely to a distance of the half the level-set length-scale or two elements. Unfortunately, the numerical fill times are difficult to obtain accurately as the gas phase viscosity makes a large difference in how fast a simulation will fill for the same value of pressure. For instance, when we reduced the second phase viscosity by a factor of 10 we got an increase in the inflow velocity when keeping all other parameters constant. Thus, the fact that the gas phase is harder to push out than it is for the experiments, adds a great deal of uncertainty to the fill times. However, we hope to be able to predict trends. Figure 12 shows the effect of the distributor geometry on the meniscus shape for Mesh 1, Mesh 2, and Mesh 3 in a vertical orientation.

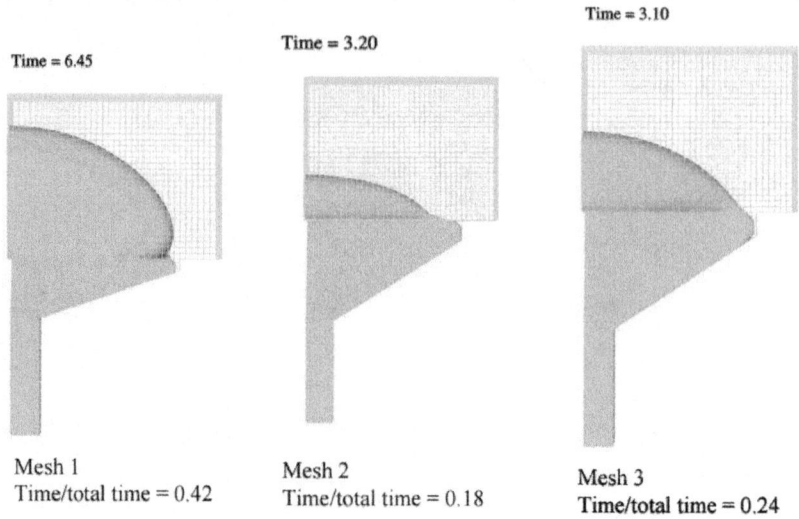

Figure 12: Free surface profile after filling the distributor for Mesh 1, Mesh 2, and Mesh 3 for vertical mould orientation.

Comparing Figure 12 and Figure 5, the numerical and experimental version of this profile, we can see that the simulations are exhibiting the physically correct trends. The original mesh takes the longest fractional time to fill the distributor, 42%, and gives the most bulging front shape. Mesh 2 is an improvement, taking 18% of the time to fill the distributor, while Mesh 3 is somewhere in between at 24%. The values for the experimental distributor dimensionless fill times are 32%, 13%, and 24%, so the simulations are also capturing the correct trends for fill time though they are not quantitative. The shape of the meniscus for Mesh 1 has more of a bulge at the edge of the distributor than the experimental meniscus, which looks as if the front is pinned at the distributor. We can also look at the profiles and dimensionless time to hit the back wall. These results are given in Figure 13.

Comparing Figure 13 to Figure 6, for the simulation versus experiment, we can see differences in the meniscus shape. The numerical interface reaches the back wall for Mesh 1 and Mesh 3, before it wets the sidewall and Mesh 2 has a flatter profile in the experiments than the simulations. The percentage time to reach the back wall for the simulations on Mesh 1, Mesh 2, and Mesh 3 are 60%, 70%, and 55% compared to 80%, 86% and 82% for the experiments. Again, we capture the correct trends, but are still unable to match the data quantitatively. Figure 14 shows the full meniscus shape and void locations for the simulations on Mesh 1, Mesh 2, and Mesh 3.

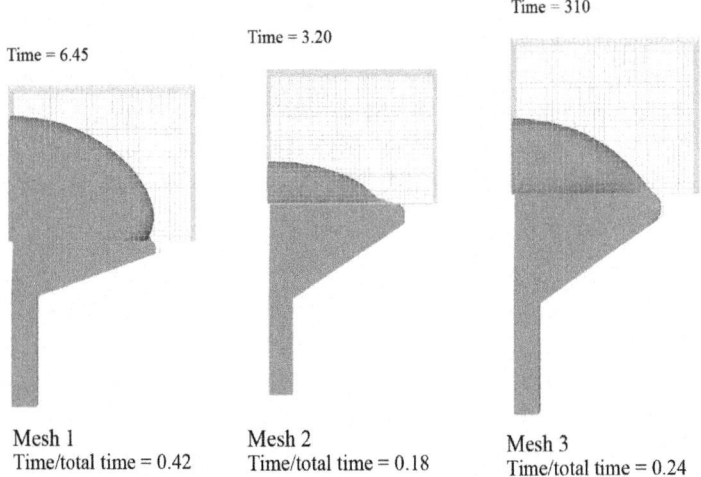

Time = 310

Time = 3.20

Time = 6.45

Mesh 1
Time/total time = 0.42

Mesh 2
Time/total time = 0.18

Mesh 3
Time/total time = 0.24

Figure 13: Free surface profile after hitting the back wall for Mesh 1, Mesh 2, and Mesh 3 for vertical mould orientation.

The void in the corner near the distributor does eventually fill in, since its size is less than the level-set length scale and we are using a diffuse interface method. The larger void at the vent never fills in as the viscous gas phase is trapped away from the vent by the fluid. For the numerical solutions it is hard to make any predictions about void size, though we can say that for similar values of the dimensionless time the voids for Mesh 2 will be smaller than Mesh 1, with Mesh 3 being somewhere in between, which does follow the experimental trend.

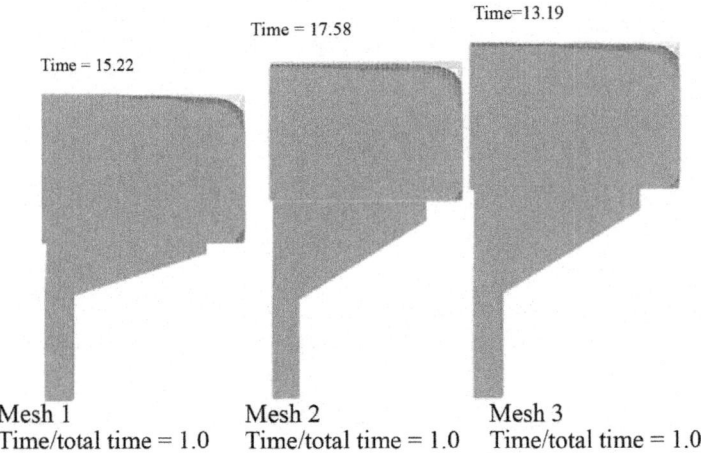

Time=13.19

Time = 17.58

Time = 15.22

Mesh 1
Time/total time = 1.0

Mesh 2
Time/total time = 1.0

Mesh 3
Time/total time = 1.0

Figure 14: Final void location and front profile for Mesh 1, Mesh 2, and Mesh 3.

We also examined the horizontal orientation numerically, which gave fill times for Mesh 1, Mesh 2, and Mesh 3 of 21.4s, 23.1s, and 22.2s compared to 23.3s, 28.9s, and 27.3s for the experiments. Again, we follow the trends of the experiment, but do not match quantitatively. Mesh 2 seems to take a longer time to fill for the experiment than one would predict numerically. Figure 15 shows a comparison of the profiles for Mesh 1 in a vertical and horizontal orientation.

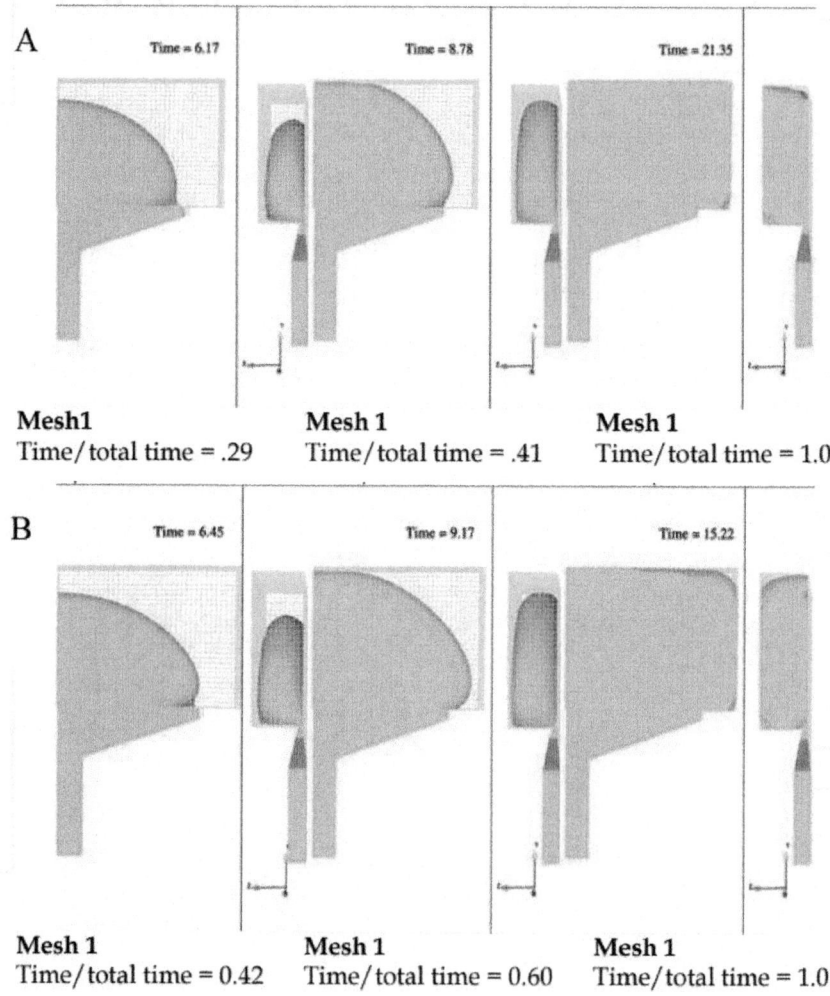

Figure 15: Mesh 1 oriented horizontally (A) and vertically (B). Leftmost pictures show profiles when the distributor is filled, middle shows profile when the fluid hits the back wall, and rightmost pictures show final profile. Both front and side views are given to highlight void location.

Comparing Figure 15 to the experimental equivalent, Figure 8, we can see that we have quantitative differences but do match some trends. The numerical meniscus shape leaving the distributor looks similar to the experimental profile as it bulges more in the centre for the horizontal orientation, though the simulation is less dramatic. The numerical profiles when the fluid first hits the back wall are flatter for the vertical orientation than the horizontal, though the vertical should be even flatter to match the data. The numerical solutions predict two voids for each orientation, though the experiments do not show a second void for the horizontal orientation near the vent. However, this void may just be difficult to see experimentally. Conversely, the horizontal void at the outflow may be an artefact of the numerical method as we have a difficult balance at the outflow between wetting forces, gravity, the gas phase viscosity, and the material flowing out the vent. Also, our numerical vent is not identical to the experimental one and exhibits a slightly different area and shape. Figure 16 shows the meniscus profiles for Mesh 1, Mesh 2, and Mesh 3 after filling the distributor for the horizontal orientation.

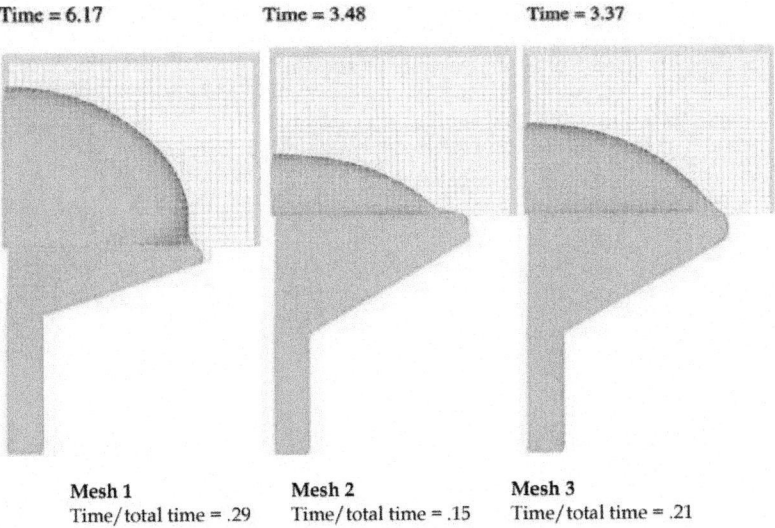

Figure 16: Free surface profile after filling the distributor for Mesh 1, Mesh 2, and Mesh 3 for horizontal mould orientation.

Comparing Figure 16 and Figure 9, the numerical and experimental version of this profile, we can see that the simulations are again exhibiting the correct trends of the physical situation. Mesh 1 shows the most pooling at the centre of the mould, Mesh 2 has a flatter profile as does Mesh 3. The experiments predict filling times for Mesh 3 to be in between Mesh 1 and

Mesh 2, and the simulations follow this trend. The dimensionless times to fill the distributor numerically for Mesh 1, Mesh 2, and Mesh 3 are 29%, 15%, and 21% compared to 26%, 13% and 22% for the experiments. In general, the simulations predict the vertical filling to be faster overall than the horizontal by several seconds for each of the geometries, whereas the experiments are faster in the vertical for Mesh 2 and 3, but slower for Mesh 1. This could have resulted from some experimental errors or from the uncertainty in injection rates and flow profiles from the experiments to the simulations.

Figure 17 shows the free surface profile as the fluid hits the back wall for Mesh 1, Mesh 2, and Mesh 3 in the horizontal orientation. The dimensionless times it takes to hit the back wall for Mesh 1, Mesh 2, and Mesh 3 are 41%, 55%, and 52% compared to values of 82%, 71%, and 70% for the experiments seen in Figure 10.

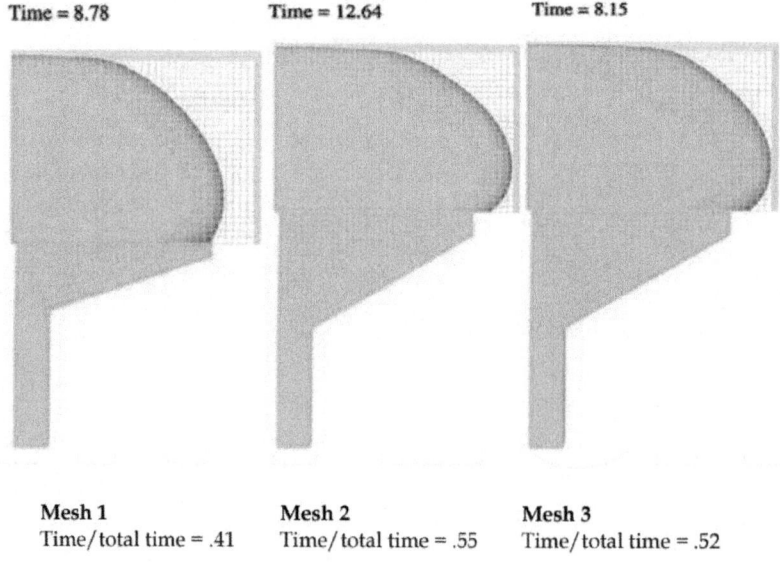

| Time = 8.78 | Time = 12.64 | Time = 8.15 |

| Mesh 1 | Mesh 2 | Mesh 3 |
| Time/total time = .41 | Time/total time = .55 | Time/total time = .52 |

Figure 17: Free surface profile after hitting the back wall for Mesh 1, Mesh 2, and Mesh 3 for horizontal mould orientation.

For the simulations, Mesh 1 hits the back wall for the smallest dimensionless time while Mesh 2 and Mesh 3 take about the same time. For the experiments, Mesh 1 takes the longest time, while Mesh 2 and Mesh 3 do take about the same time. Thus for this case, we are capturing one trend, but not the differences between Mesh 1 and Mesh 2. Table 4 summarizes the fill times to reach the distributor, back wall, and complete mould for the simulations and experiments for vertical and horizontal orientations on all three meshes.

Table 4: Summary of fill times for experiment and simulations to reach the distributor, the wall and completely full

Mesh	Orientation	Expt. Time - Full	Expt. % Time - Dist	Expt. % Time - Wall	Sim. Time- Full	Sim. % Time - Dist	Sim. % Time - Wall
1	Vertical	24.6s	32%	80%	15.2s	42%	60%
2	Vertical	26.9s	13%	86%	17.5s	18%	70%
3	Vertical	24.6s	24%	82%	13.2s	24%	55%
1	Horizontal	23.3s	26%	82%	21.4s	29%	41%
2	Horizontal	28.9s	13%	71%	23.1s	15%	55%
3	Horizontal	27.3s	22%	71%	15.6s	21%	52%

From Table 4 we can see that we capture some of the correct trends, especially for the vertical orientation, though some features elude us like the time to fill to the back wall for Mesh 1 in the horizontal orientation. Sources of uncertainty in the simulations include: 1) Lack of clarity of inflow conditions from the experiment to the simulation, since it is somewhere in between constant pressure and constant velocity, 2) Possible poor performance of the Blake wetting model at moderate capillary number and large dynamic contact angles, 3) Possible poor performance of the wetting model at wetting speed higher than experiments used to populate the model. Sources of error in the simulations include: 1) Numerically expedient of high gas phase viscosity, 2) Lack of compressibility for the gas dynamics, 3) The use of a diffuse interface model that smears out material property jumps, allowing viscous bleed through of the liquid phase into the gas phase. For future work, we will try to reduce these uncertainties and errors by using some of the advanced features in ARIA, which should be available soon, to allow for smaller values of the gas phase viscosity, such as sharp integration and a compressible gas phase. The optimal choice of wetting model for moderate to high capillary numbers continues to be an ongoing focus of our research.

CONCLUSION

A diffuse interface finite element/level-set algorithm has been used to investigate filling behaviour for injection loading using a Blake wetting model. The modelling has been successful in matching experimental data qualitatively, but quantitative agreement is still lacking especially for the wetting dynamics and meniscus shape. For future work, we will investigate an advanced version of the level set method termed the conformal decomposition finite element method (CDFEM). CDFEM is a hybrid moving boundary algorithm, which uses a level set field to determine the location of the fluid-fluid interface and then dynamically adds mesh on the interface to facilitate the resolution of discontinuous material properties and fields, as well as the application of boundary conditions such as capillarity. This is a sharp interface method, where it is possible to apply jumps in material properties, material models, and field variables [Noble et al, 2010]. We believe this algorithm, which should be available soon, will lead to better agreement with experiments and should allow for straightforward inclusion of a compressible gas phase.

ACKNOWLEDGMENTS

This work was supported by the Laboratory Directed Research and Development program at Sandia National Laboratories. We would like to thank Dr. Pin Yang, the project principal investigator, for his support of this study. We would also like to thank our Sandia reviewer Daniel Guildenbecher and P. Randall Schunk for their insightful editorial comments.

Sandia National Laboratories is a multi-program laboratory managed and operated by Sandia Corporation, a wholly owned subsidiary of Lockheed Martin Corporation, for the U.S. Department of Energy's National Nuclear Security Administration under contract DEAC04-94AL85000.

REFERENCES

1. Bach, P. & Hassager, O. (1985). "An algorithm for the use of the Lagrangian specification in Newtonian fluid mechanics and applications to free surface flow," J. Fluid Mech., 152, 173-190.

2. Blake, T. D. & Haynes, J. M. (1969). "Kinetics of liquid/liquid displacement," J. Colloid Interface Sci., 30, 421.

3. Blake, T. D. & De Coninck, J. (2002). "The Influence of Solid-Liquid Interactions on Dynamic Wetting," Adv. Colloid & Int. Sci., 96, 21-36.

4. Blake, T. D. (2006). "The physics of moving wetting lines," J. Colloid Interface Sci., 299, 1-13.

5. Brackbill, J. U.; Kothe, D. B. & Zemach, C. (1992). "A continuum method for modeling surface-tension," J. Comp. Phys., 100, 335-354.

6. Brooks, C. F.; Grillet, A. M. & Emerson, J. A. (2006). "Experimental investigation of the spontaneous wetting of polymers and polymer blends," Langmuir, 22, 9928-9941.

7. Cox, R. G. (1986). "The dynamics of the spreading of liquids on a solid surface. Part 1. Viscous flow," J. Fluid Mech., 168, 169-194.

8. Dohrmann, C. R. & Bochev, P. B. (2004). "A stabilized finite element method for the Stokes problem based on polynomial pressure projections,"

Int. J. Num. Meth. Fluids, 46, 183–201.

9. Donea, J. (1984). "A Taylor-Galerkin Method for Convective Transport Problems," Int. J. Num. Meth. Engn., 20, 101-119.

10. Hirt, C. W. & Nichols, B. D. (1981). "Volume of fluid (VOF) method for the dynamics of free boundaries," J. Comp. Phys., 39, 201-225.

11. Hoffman, R. L. (1975). "A study of the advancing interface. I. Interface shape in liquid-gas systems," J. Colloid Inter. Sci., 50, 228-241.

12. Hwang, C. J. & Kwon, T. H. (2002). "A full 3D finite element analysis of the powder injection moulding filling process including slip phenomena," Poly. Eng. Sci, 42, 33-50.

13. Hughes, T. J. R. (2000). The Finite Element Method. Dover Publications, New York, USA.

14. Kauzlaric, D.; Pastewka, L.; Meyer, H.; et al., (2011). "Smoothed particle hydrodynamics simulation of shear-induced powder migration in injection moulding," Phil. Trans. Roy. Soc. A Math. Phys. Engn. Sci., 369, 2320-2328.

15. Kistler, S. F. (1983). "The fluid mechanics of curtain coating and related viscous free surface flows with contact lines," Ph.D. Thesis, University of Minnesota, Minneapolis.

16. Kulasegaram, S.; Bonet, J.; Lewis, R. W. & Profit, M., (2003). "High pressure die casting simulation using a Lagrangian particle method," Comm. Appl. Numer. Meth. Engn, 19, 679-687.

17. Kumar, A. & Ghoshdastidar, P. S. (2002). "Numerical Simulation of Polymer Flow in a Cylindrical Cavity," J. Fluids Engn., 124, 251-261.

18. Lewis, R. W.; Navti, S. E. & Taylor, C. (1997). "A mixed Lagrangian-Eulerian approach to modelling fluid flow during mould filling," Int. J. Num. Meth. Fluids, 25, 931-952.

19. Love, E. & Sulsky, D. L. (2006). "An energy-consistent material-point method for dynamic finite deformation plasticity," Int. J. Num. Meth. Engn., 65, 1608-1638.

20. Ilinca, F. & Hetu, J. -F. (2000). "Finite element solution of three-dimensional turbulent flows applied to mould-filling problems," Int. J. Num. Meth. Fluids, 34, 729-750.

21. Ilinca, F. & Hetu, J.-F. (2001). "Three-dimensional filling and post-filling simulation of polymer injection moulding," Int. Poly. Proc., 16, 291-301.

22. Ilinca, F. & Hetu, J.-F. (2008). "Three-dimensional free surface flow simulation of segregating dense suspensions ," Int. J. Num. Meth. Fluids, 58, 451-472.

23. Mondy, L.A.; Rao, R. R.; Brooks, C. F.; Noble, D. R.; et al., (June 2007). "Wetting and free surface flow modelling for potting and encapsulation," SAND2007-3316, Sandia National Laboratories, Albuquerque, NM.

24. Nithiarasu, P. (2005). "An arbitrary Lagrangian Eulerian (ALE) formulation for free surface flows suing a characteristic-base split (CBS) scheme," Int. J. Num. Meth. Fluids, 48, 1415-1428.

25. Noble, D. R.; Newren, E. & Lechman, J. B. (2010). "A conformal decomposition finite element method for modelling stationary fluid interface problems", Int. J. Num. Meth. Fluids, 63, 725-742.

26. Notz, P. K.; Subia, S. R.; Hopkins, M. M.; Moffat, H. K. & Noble, D. R. (April 2007). "ARIA Manual Aria 1.5: User's Manual," SAND2007-2734, Sandia National Laboratories, Albuquerque, NM.

27. Oishi, C. M.; Tome, M. F.; Cuminato, J. A. & McKee, S. (2008). "An implicit technique for solving 3D low Reynolds number moving free surface flows, J. Comp. Phys., 227, 7446-7468.

28. Radovitzky, R. & Ortiz, M. (1998). "Lagrangian finite element analysis of Newtonian fluid flows, Int. J. Num. Meth. Engn., 43, 607-617.

29. Rao, R. R.; Mondy, L. A.; Noble, D. R.; Hopkins, M. M.; Notz, P. K.; Baer, T. A.; Halbleib, L.; Yang, P.; Burns, G.; Grillet, A. M.; Brooks, C.; Cote, R. O. & Castaneda, J. N. (September 2006). "Modeling Injection Molding of Net-Shape Active Ceramic Components," SAND2006-6786, Sandia National Laboratories, Albuquerque, NM.

30. Rao, R. R.; Mondy, L. A.; Noble, D. R.; Moffat, H. K; Adolf, D. B. & Notz, P.K. (2011). "A Level Set Method to Study Foam Processing: A Validation Study," Int. J. Num. Meth. Fluids, early view.

31. Ren, W.; Hu, D. & E, W. (2010). "Continuum models for the contact line problem," Phys. Fluids, 22.

32. Roberts, S. A. & Rao, R. R. (2011). "Entraining flow of a shear-thinning jet impinging in a container: A finite element approach," J. Non-Newtonian Fluid Mech., 166, 1100– 1115.

33. Sackinger, P. A.; Schunk, P. R. & Rao, R. R. (1996). "A Newton-Raphson pseudo-solid domain mapping technique for free and moving boundary problems: a finite element implementation," J. Comp. Phys, 125, 83-103.

34. Sartor, L. (1990). "Slot Coating: Fluid Mechanics and Die Design," Ph.D. Thesis, University of Minnesota, Minneapolis.

35. Schunk, P. R.; Noble, D. R.; Baer, T. A.; Secor, R. B. & Jendoubi, S. (September 2006). "Implementation and performance of published wetting models in level-set and ALE-based algorithms for free and

moving boundary problems," Poster Presentation, 13th International Coating Science and Technology Symposium, Denver, Colorado.

36.   Sethian, J. A. (1999). Level Set Methods and Fast Marching Methods, Volume 3 of Cambridge Monographs on Applied and Computational Mathematics. Cambridge University Press, New York, USA, 2nd edition.

37.   Shikhmurzaev, Y. D. (1994). "Mathematical modelling of wetting hydrodynamics," Fluid Dynamics Res., 13, 45-64.

38.   Tome, M. F.; Filho, A. C.; Cuminato, J. A.; Mangiavacchi, N. & McKee, S. (2001). "GENSMAC3D: a numerical method for solving unsteady three-dimensional free surface flows," Int. J. Num. Meth. Fluids, 37, 747-796.

39.   Ville, L.; Silva, L.; & Coupez, T. (2011) "Convected level set method for the numerical simulation of fluid buckling," Int. J. Num. Meth. Fluids, 66, 324-344. Voinov, O. V. (1976). "Hydrodynamics of wetting," Fluid Dynamics, 11, 714-721.

40.   Zhang, J. & Khayat, R. E. (2001). "A Lagrangian boundary element approach to transient three-dimensional free surface flow in thin cavities, Int. J. Num. Meth. Fluids, 37, 399-418.

Chapter 2

EXPERIMENTALLY VALIDATED NUMERICAL MODELING OF HEAT TRANSFER IN GRANULAR FLOW IN ROTATING VESSELS

Bodhisattwa Chaudhuri[1], Fernando J. Muzzio[2] and M. Silvina Tomassone[2]

[1]Department of Pharmaceutical Sciences, University of Connecticut, Storrs, CT, 06269

[2]Department of Chemical and Biochemical Engineering, Rutgers University, Piscataway, NJ, 08854 United States of America

INTRODUCTION

Heat transfer in particulate materials is a ubiquitous phenomenon in nature, affecting a great number of applications ranging from multi-phase reactors to kilns and calciners. The materials used in these type of applications are typically handled and stored in granular form, such as catalyst particles, coal, plastic pellets, metal ores, food products, mineral concentrates, detergents, fertilizers and many other dry and wet chemicals. Oftentimes, these materials need to be heated and cooled prior to or during processing. Rotary calciners are most commonly used mixing devices used in metallurgical and catalyst industries (Lee, 1984; Lekhal et. al., 2001). They are long and nearly horizontal rotating drums that can be equipped with internal flights (baffles) to process various types of feedstock. Double cone impregnators are utilized to incorporate metals or other components into porous carrier particles while developing supported catalysts. Subsequently, the impregnated catalysts are heated, dried and reacted in rotating calciners to achieve the desired final form. In these processes, heat is generally transferred by conduction and convection between a solid surface and particles that move relative to the surface.

Over the last fifty years, there has been a continued interest in the role of system parameters and in the mechanisms of heat transfer between granular media and the boundary surfaces in fluidized beds (Mickey & Fairbanks, 1955; Basakov, 1964; Zeigler & Agarwal, 1969; Leong et.al., 2001; Barletta et. al., 2005), dense phase chutes, hoppers and packed beds (Schotte, 1960; Sullivan & Sabersky, 1975; Broughton & Kubie, 1976; Spelt et. al., 1982; Patton et.

al., 1987; Buonanno & Carotenuto, 1996; Thomas et. al., 1998; Cheng et. al., 1999), dryers and rotary reactors and kilns (Wes et. al., 1976; Lehmberg et. al., 1977). More recently, experimental work on fluidized bed calciner and rotary calciners/kilns have been reported by LePage et.al, 1998; Spurling et.al., 2000, and Sudah. et al., 2002. In many of these studies, empirical correlations relating bed temperature to surface heat transfer coefficients for a range of operating variables have been proposed. Such correlations are of restricted validity because they cannot be easily generalized to different equipment geometries and it is risky to extrapolate their use outside the experimental range of variables studied.

Moreover, most of these models do not capture particle-surface interactions or the detailed microstructure of the granular bed. Since the early 1980s, several numerical approaches have been used to model granular heat transfer methods using (i) kinetic theory (Natarajan & Hunt, 1996) (ii) continuum approaches (Michaelides, 1986; Ferron & Singh, 1991; Cook & Cundy, 1995, Natarajan & Hunt, 1996, Hunt, 1997) and (iii) discrete element modeling (DEM) (Kaneko et. al., 1999; Li & Mason, 2000; Vargas & McCarthy, 2001; Skuratovsky et. al., 2005). The constitutive model based on kinetic theory incorporates assumptions such as isotropic radial distribution function, a continuum approximation and purely collisional interactions amongst particles, which are not completely appropriate in the context of actual granular flow. Continuum models neglect the discrete nature of the particles and assume a continuous variation of matter that obeys the laws of conservation of mass and momentum. To the best of our knowledge, among continuum approaches, only Cook and Cundy, 1995 modeled heat transfer of a moist granular bed inside a rotating vessel. Continuum-based models can yield accurate results for the time-averaged quantities such as velocity, density and temperature while simulating heat transfer in granular material, but fail to reveal the behavior of individual particles and do not consider inter-particle interactions.

In the discrete element model, each constituent particle is considered to be distinct. DEM explicitly considers inter-particle and particle-boundary interactions, providing an effective tool to solve the transient heat transfer equations. Most of the DEM-based heat transfer work has been either two-dimensional or in static granular beds. To the best of our knowledge no previous work has used three-dimensional DEM to study heat transfer in granular materials in rotary calciners (with flights attached) that are the subject of this study. Moreover, a laboratory scale rotary calciner is used to estimate the effect of various materials and system parameters on heat transfer, which also helps to validate the numerical predictions.

EXPERIMENTAL SETUP

A cylindrical tubing (8 inches outer diameter, 6 inches inner diameter and 3 inches long) of aluminum is used as the "calciner" for our experiments. The calciner rides on two thick Teflon wheels (10 inches diameter) placed at the two ends of the calciner, precluding the direct contact of the metal wall with the rollers used for rotating the calciner. The side and the lateral views of the calciner are shown in Figure 1a and 1b respectively. Figure 1a also shows how the ten thermocouples are inserted vertically into the calciner with their positions being secured at a constant relative position (within themselves) using a rectangular aluminum bar attached to the outer Teflon wall of the calciner. Twelve holes are made on the Teflon wall of the calciner where the two holes at the end are used to secure the aluminum bar with screws, whereas, the intermediate holes allow the insertion of 10 thermocouples (as shown in Fig 1c). The other end-wall of Teflon has a thick glass window embedded for viewing purpose. In Figure 1d, the internals of the calciner comprising the vertical alignment of 10 thermocouples is visible through the glass window.

Thermocouples are arranged radially due to the radial variation of temperature during heat transfer in the granular bed as observed in our earlier simulations (Chaudhuri et.al, 2006). The thermocouples are connected to the Omega 10 channel datalogger that works in unison with the data acquisition software of the adjacent PC. 200μm size alumina powder and cylindrical silica particles (2mm diameter and 3mm long) are the materials used in our experiments. The calciner is initially loaded with the material of interest. Twenty to fifty percent of the drum is filled with granular material during the experiments. At room temperature, an industrial heat gun is used to uniformly heat the external wall of the calciner. The calciner is rotated using step motor controlled rollers, while the wall temperature is maintained at 100°C. At prescribed intervals, the "calciner" is stopped to insert the thermocouples inside the granular bed to take the temperature readings. Once temperature is recorded, the thermocouples are extracted and rotation is initiated again.

NUMERICAL MODEL AND PARAMETER USED

The Discrete Element Method (DEM), originally developed by Cundall and Strack (1971, 1979), has been used successfully to simulate chute flow (Dippel, et.al., 1996), heap formation (Luding, 1997), hopper discharge (Thompson and Grest, 1991; Ristow and Hermann, 1994), blender segregation (Wightman, et.al, 1998; Shinbrot, 1999; Moakher, 2000) and flows in rotating drums (Ristow, 1996; Wightman, et.al., 1998). In the present study DEM is used to simulate the dynamic behavior of cohesive and non-cohesive powder in a rotating drum (calciner) and double cone (impregnator). Granular material is considered here as a collection of frictional inelastic spherical particles. Each particle may interact with its neighbors or with the boundary only at contact points through normal and tangential forces. The forces and torques acting on each of the particles are calculated as:

$$\sum F_i = m_i g + F_n + F_t + F_{cohes}$$

(1)

$$\sum T_i = r_i \times F_T$$

(2)

Thus, the force on each particle is given by the sum of gravitational, inter-particle (normal and tangential: F_N and F_T) and cohesive forces as indicated in Eq. (1). The corresponding torque on each particle is the sum of the moment of the tangential forces (F_T) arising from inter-particle contacts (Eq. (2)). We use the "latching spring model" to calculate normal forces. This model, developed by Walton and Braun (1986, 1992, 1993), allows colliding particles to overlap slightly. The normal interaction force is a function of the overlap. The normal forces between pairs of particles in contact are defined using a spring with constants K_1 and K_2: $F_N = K_1 \alpha_1$ (for compression), and $F_N = K_2 (\alpha_1 - \alpha_0)$ (for recovery). These spring constants are chosen to be large enough to ensure that the overlaps α_1 and α_0 remain small compared to the particles sizes. The degree of inelasticity of collisions is incorporated in this model by including a coefficient of restitution $e = (K_1/K_2)^{1/2}$ ($0 < e = 1$ implies perfectly elastic collision with no energy dissipation and $e=0$ implies completely inelastic collision).

Tangential forces (F_T) in inter-particle or particle-wall collision are calculated with Walton's incrementally slipping model. After contact occurs, tangential forces build up, causing displacement in the tangential plane of contact. These forces are assumed to obey Coulomb's law. The initial tangential stiffness is considered to be proportional to the normal stiffness. If the magnitude of tangential forces is greater than the product of the normal force by the coefficient of static friction, (i.e. $T \geq \mu FN$) sliding takes place with a constant coefficient of dynamic friction. The model also takes into account the elastic deformation that can occur in the tangential direction. The tangential force T is evaluated considering an effective tangential stiffness k_τ associated with a linear spring. It is incremented at each time step as $T_{t+1} = T_t + k_t \Delta s$, where Δs is the relative tangential displacement between two time steps (for details on the definition of Δs see Walton (1993)). The described model was used successfully to perform three-dimensional simulations of granular flow in realistic blender geometries, where it confirmed important experimental observations (Wightman, et.al., 1998, Moakher, et. al., 2000, Shinbrot, et.al., 1999; Sudah, et.al., 2005).

Figure 1: (a) Aluminum calciner on rollers (side view) showing 10 thermocouples inserted within the calciner through the Teflon side-wall. (b) Lateral view of the calciner. (c) 10 thermocouples are tied up to the metal rod which is being attached to the teflon wall. Vertically located, ten holes are also shown in the teflon wall through which thermocouples are inserted inside the calciner. (d) Another side view showing the internals of the calciner and the vertical alignment of 10 thermocouples which are visible through the glass window.

We also incorporate cohesive forces between particles in our model using a square-well potential. In order to compare simulations considering different numbers of particles, the magnitude of the force was represented in terms of the dimensionless parameter $K = F_{cohes}/mg^1$, where K is called the bond number and is a measure of cohesiveness that is independent of particle size, F_{cohes} is the cohesive force between particles, and mg is the weight of the particles. Notice that this constant force may represent short range effects2 such as electrostatic or van der Waals forces. In this model, the cohesive force (F_{cohes}) between two particles or between a particle and the wall is unambiguously defined in terms of K. Four friction coefficients need to be defined: particle-particle and particle-wall static and dynamic coefficients. Interestingly, (and unexpectedly to the authors) all four friction coefficients turn out to be important to the transport processes.

Heat transport within the granular bed may take place by: thermal conduction within the solid; thermal conduction through the contact area between two particles in contact; thermal conduction through the interstitial fluid; heat transfer by fluid convection; radiation heat transfer between the surfaces of particles. Our work is focused on the first two mechanisms of conduction which are expected to dominate when the interstitial medium is stagnant and composed of a material whose thermal conductivity is small compared to that of the particles. O'Brien (1977) estimated this assumption to be valid as long as ($k_S a / k_f r$) >> 1), where a is the contact radius, r is the particle radius of curvature, k_f denotes the fluid interstitial medium conductivity and k_S is the thermal conductivity of the solid granular material. This condition is identically true when k_f=0, that is in vacuum. Heat transport processes are simulated accounting for initial material temperature, wall temperature, granular heat capacity, granular heat transfer coefficient, and granular flow properties (cohesion and friction).

Heat transfer is simulated using a linear model, where the flux of heat transported across the mutual boundary between two particles i and j in contact is described as

$$Q_{ij} = H_c(T_j - T_i)$$

(3)

Here. T_i and T_j are the temperatures of the two particles and the inter-particle conductance H_c is:

$$H_c = 2k_S \left[\frac{3F_N r^*}{4E^*} \right]^{1/3}$$

(4)

where k_S is the thermal conductivity of the solid material, E* is the effective Young's modulus for the two particles, and r* is the geometric mean of the particle radii (from Hertz's elastic contact theory). The evolution of temperature of particle i from its neighbor (j) is

$$\frac{dT_i}{dt} = \frac{Q_i}{\rho_i C_i V_i}$$

(5)

Here, Q_i is the sum of all heat fluxes involving particle i and $\rho_i C_i V_i$ is the thermal capacity of particle i.

Equations (3-5) can be used to predict the evolution of each particle's temperature for a flowing granular system in contact with hot or cold surfaces. The algorithm is used to examine the evolution of the particle temperature both in the calciner and the double cone impregnator. This numerical model is developed based on following assumptions:

- Interstitial gas is neglected.
- Physical properties such a heat capacity, thermal conductivity and Young Modulus are considered to be constant.
- During each simulation time step, temperature is uniform in each particle (Biot Number well below unity).
- Boundary wall temperature remains constant.

The major computational tasks at each time step are as follows: (i) add/delete contact between particles, thus updating neighbor lists, (ii) compute contact forces from contact properties, (iii) compute heat flux using thermal properties (iv) sum all forces and heat fluxes on particles and update particle position and temperatures, and (v) determine the trajectory of the particle by integrating Newton's laws of motion (second order scalar equations in three dimensions). A central difference scheme, Verlet's Leap Frog method, is used here.

The computational conditions and physical parameters considered are summarized in Table 1. Heat transport in alumina is simulated for the experimental validation work, and then copper is chosen as the material of interest for further investigation on baffle size/orientation in calciners and impregnators.We simulated the flow and heat transfer of 20,000 particles of 1mm size rotated in the calciner equipped with or without baffle of variable shapes. The calciner consists of a cylindrical 6 inch diameter vessel with length of 0.6 inches, intentionally flanked with frictionless side walls to simulate a thin slice of the real calciner, devoid of end-wall effects. Two baffle sizes are considered (of thicknesses equal to 3cm and 6cm). The initial surface temperature of all the particles is considered to be 298 K (room temperature) whereas the temperature of the wall (and the baffle in the impregnator) is considered to be constant, uniform, and equal to 1298 K. The computational conditions and physical parameters considered are summarized in Table 1. Initially particles were loaded into the system and allowed to reach mechanical equilibrium. Subsequently, the temperature of the vessel was suddenly raised to a desired value, and the evolution of the temperature of each particle in the system was recorded as a function of time.

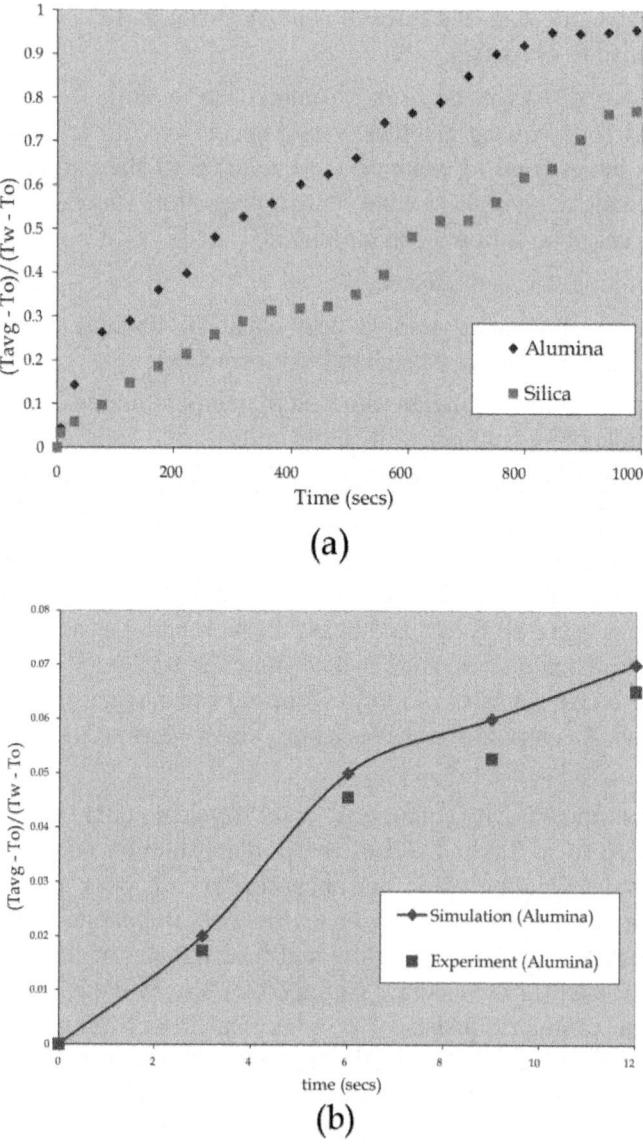

Figure 2: (a) Variation of average bed temperature with time for alumina and silica; (b) Evolution of average bed temperature for simulation and experiments with alumina. The fill level of the calciner is 50% and is rotated at 20rpm in the experiments and simulations

The double cone impregnator model considers flow and heat transfer of 18,000 particles of 3mm diameter in a vessel of 25 cm diameter and 30 cm length. The cylindrical portion of the impregnator is 25 cm diameter and 7.5 cm long. Each of the conical portions is 11.25 cm long and makes an angle of 45° with the vertical axis. The diameter at the top or bottom of the impregnator is 2.5cm the effect of baffle size is investigated in impregnators. Intuitively, the baffle is kept at an angle 45° with respect to the axis of rotation. The length of the baffle is 25cm, same as the diameter of the cylindrical portion of the impregnator. The width and thickness of the baffle are equal to one another (square cross section). In order to describe quantitatively the dynamics of evolution of the granular temperature field, the following quantities were computed:

- Particle temperature fields vs. time
- Average bed temperature vs. time
- Variance of particle temperatures vs. time

These variables were examined as a function of relevant parameters, and used to examine heat transport mechanisms in both of the systems of interest here

RESULTS AND DISCUSSIONS

Effect of thermal properties in calciners

The effect of thermal conductivity in heat transfer is examined using alumina and silica particles separately, each occupying 50% of the calciner volume. The calciner is rotated at the speed of 20 rpm. The average bed temperature (T_{avg}) is estimated as the mean of the readings of the ten thermocouples and scaled with the average wall temperature (Tw) and the average initial condition (To) of the particle bed to quantify the effect of thermal conductivity. In Figure 2a, as expected, alumina with higher thermal conductivity warms up faster than silica. DEM simulations are performed with the same value for the physical and thermal properties of the material used in the experiments (for Alumina: thermal conductivity: k_s = 35 W/mK and heat capacity: C_p = 875 J/KgK, for Silica: K = 14 W/mK, C_p = 740 J/KgK). The initial surface temperature of all the particles is considered to be 298 K (room temperature) whereas the temperature of the wall is kept constant and equal to 398 K (in isothermal conditions). The DEM simulations predict the temperature of each of the particles in the system, thus the average bed temperature (T_{avg}) in simulation is the mean value of the predicted temperature of all the particles. Figure (2b) shows the variation of scaled average bed temperature for both simulation and experiments. The predictions of our simulation show a similar upward trend to the experimental findings.

Effect of vessel speed in the calciner

Alumina and silica powders are heated at varying vessel speed of 10, 20 and 30 rpm. The wall is heated and maintained at 100°C. Figure 3(a) and 3(b) show the evolution of average bed temperature with time as a function of vessel speed for alumina and silica respectively. The average bed temperatures for all the cases follow nearly identical trends. The external wall temperature is maintained at a constant temperature of 100°C. Figure (3c) shows the variation of scaled average bed temperature for simulation. All experimental temperature measurements were performed every 30 seconds; with a running time of 1200 seconds. However, each of our simulation runs was performed for only 12 seconds. Assuming a dispersion coefficient $E \sim \frac{L^2}{T}$ to be constant [Bird et. al., 1960; Crank, 1976], where L and T are the length and time scales, respectively, of the microscopic transitions that generate scalar transport, then the time required to achieve a certain progress of a temperature profile is proportional to the square of the transport microscale. The radial transport length scale used in the simulations, if measured in particle diameters, is much smaller than in the experiment, and correspondingly, the time scale needed to achieve a comparable progress of the temperature profile is much shorter, as presented in Figures 3a-c. In fact, the ratio of time scales between the experiment and the simulation probably is same to the ratio of length scales squared, shown by calculation below.

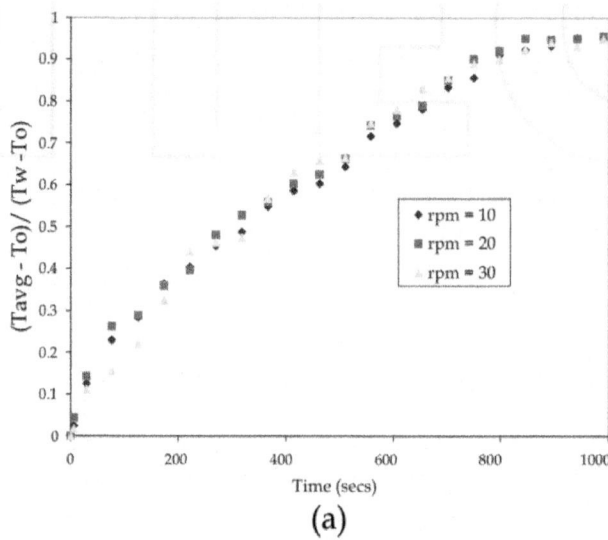

(a)

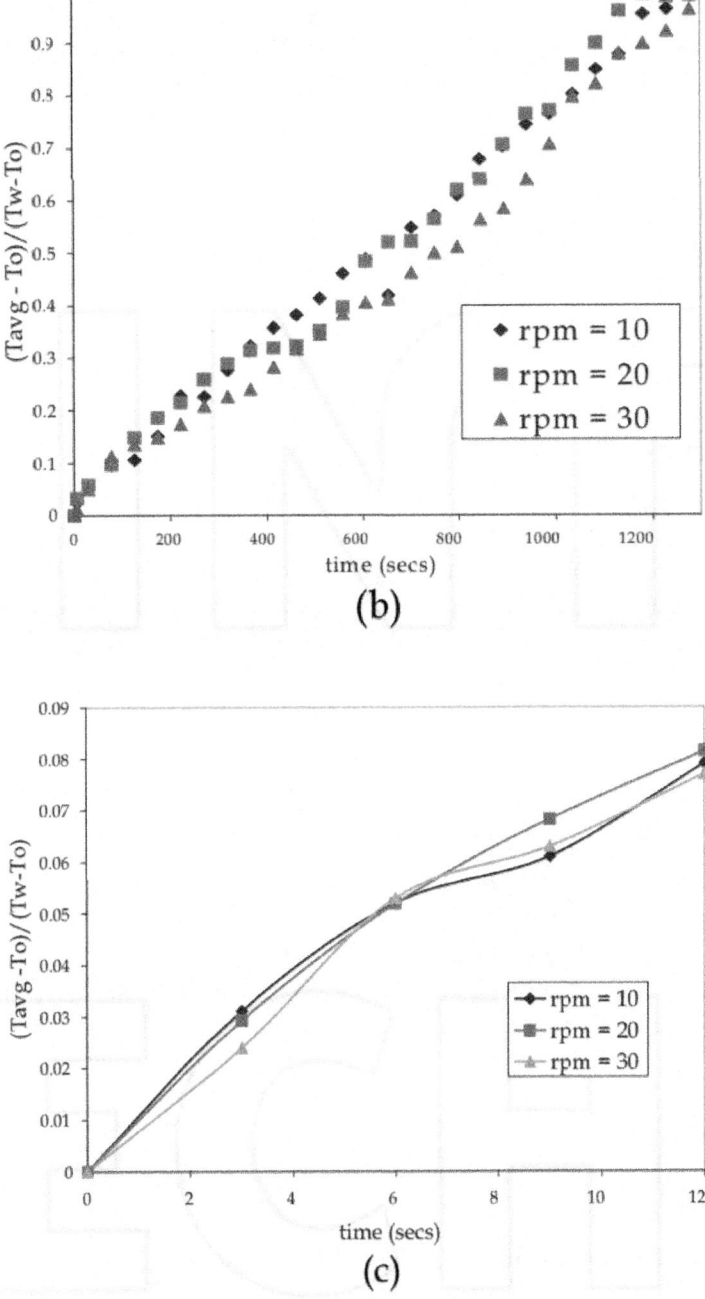

Figure 3: Variation of temperature with time as a function of vessel speed for (a) Alumina (b) silica and (c) model with alumina.

In the experiments, the diameter of the vessel (De), duration of the experiment (Te) and particle size (de) are 6 inches, 1200 seconds and 200 microns (alumina) respectively. Whereas, in the simulations, the diameter of the vessel (Ds), time of the simulation (Ts) and particle size (ds) are 6 inches, 12 seconds and 2mm respectively. Ratios of time and length scales are estimated as below:

Ratio of time scales (R_T): $\dfrac{Te}{Ts} = \dfrac{1200}{12} = 100$

Ratio of length scales (R_L): $\dfrac{L_e}{L_s} = \dfrac{\dfrac{D_e}{d_e}}{\dfrac{D_s}{d_s}} = \dfrac{\dfrac{6}{0.2}}{\dfrac{6}{2}} = 10$

Therefore, $R_T = (R_L)^2$

Although there is a big difference in the time scale in the plots of our experiments (Fig. 3a or Fig. 3b) and simulations (Fig. 3c), they still exhibit the same transport phenomena in different time scales. The predictions of our simulation show the same upward trend similar to the experimental findings, even though, they are plotted in different time scales. The nominal effect of vessel speed on heat transfer was also observed by Lybaert, 1986, in his experiments with silica sand or glass beads heated in rotary drum heat exchangers.

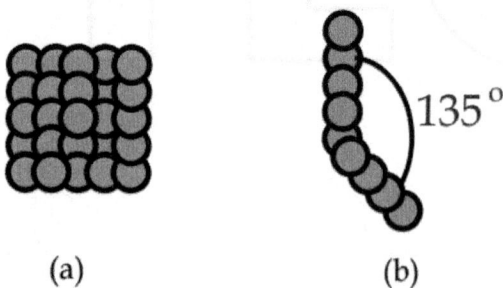

(a)   (b)

Figure 4: Baffles are formed with particles glued together (a) square cross-section and (b) L-shaped cross section.

Effect of baffles on heat transfer in calciners using a DEM model

Section 4.3 is focused on our particle simulations only. After validation of the model, presented in last two subsections, a parametric study is conducted by varying the size and the orientation of the baffles of the calciner using the same DEM model. The evolution of particle temperature is visually track using color-coding. Particles with temperature lower than 350°K are colored blue; those with temperatures between 350°K and 550°K are painted cyan; those with temperatures between 550°K and 750°K are colored green and for temperatures between 750°K and 950°K, particles are colored yellow. Particles with temperatures higher than 950°K are colored red.

Figure 5 shows a time sequence of axial snapshots of color-coded particles in the calciner. Time increases from left to right (t = 0, 3 and 9 secs), while the baffle design vary from top to bottom.

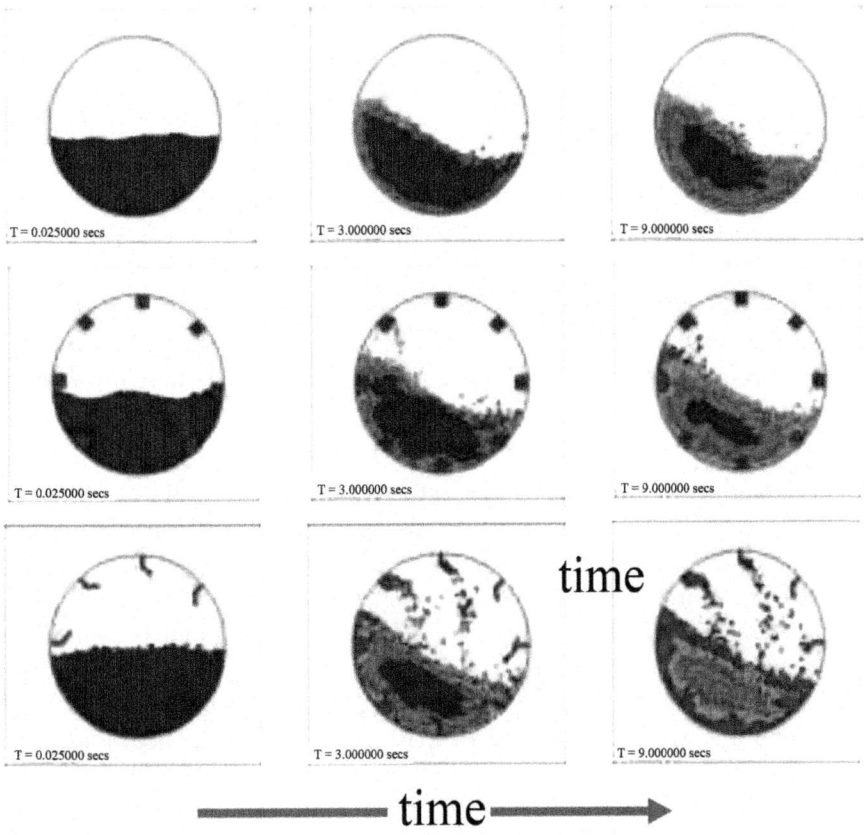

Figure 5: Time sequence of axial snapshots

Effect of Baffle Shape in Heat Transfer

In this section we study the effect of baffle shape in the calcination process. We do this by extending the DEM model of a calciner without baffles (which was previously validated) to one that which now effectively incorporates baffles.

In our model, baffles or flights are attached to the inner wall of the calciner of radius 15cm and length of 1.6cm. Baffles run longitudinally along the axial direction of the calciner. We consider 8000 copper particles of radius 2mm heated in the calciner which rotates are 20 rpm for various baffle designs. The initial temperature of the particles is chosen to be at room temperature (298°K). We simulate baffles of two different cross sections, i.e. rectangular and L-shaped by rigidly grouping particles of 2mm size, which perform solid body rotation with the calciner wall. Fig 4 depicts the composition of the different baffles. We construct the baffle particles purposely overlapping with each other by 10% of their diameter, to nullify any inter-particle gap which may cause smaller particles to percolate through the baffle. The square shaped baffle of cross sectional area of approximately 58mm2 and 340 mm2 are designed by arranging a matrix of 2 by 2 particles and 5 by 5 particles respectively. The L-shaped baffle is constructed by 9 particles bonded in a straight line until the 5th particle and then arranging the remaining 4 particles in an angle of 135°. Baffle particles also remain at the same temperature of the wall, i.e. 1298°K. For visual representation, particles are color-coded based on their temperature. In Figure 5, the axial snapshots captured at time t= 0, 1 and 3 revolutions for 3 different baffle configurations: (i) no baffle (ii) baffles of each 400 mm2 cross sectional area (iii) 8 L-shaped flights. The blue core displays the larger mass of particles at initial temperature. This cold core shrinks with time for all cases, however, the volume of the blue core shrinks faster for calciner with L shaped baffles.

The number of red particles present in the bed increases for calciners with L-shaped baffles. Thus, increased surface area of the bigger baffle enhances in heat transfer within the calciners. The effect of baffle configuration on heat transfer is quantified with our DEM model by measuring the average bed temperature as a function of time for all baffle configurations. Average bed temperature rises faster for calciners with L-shaped baffles, as seen in Figure 6(a). The uniformity of the temperature of the particle bed is quantified by estimating the standard deviation of the temperature of the bed. Figure 6(b) shows the effect of the baffle configuration on the uniformity of the bed temperature. The L-shaped baffles scoops up more particles in comparison to the square shaped baffle and helps in breaking the quasistatic zone in the center of the granular bed and redistributing the particles onto the cascading layer causing rapid mixing (uniformity) within the bed.

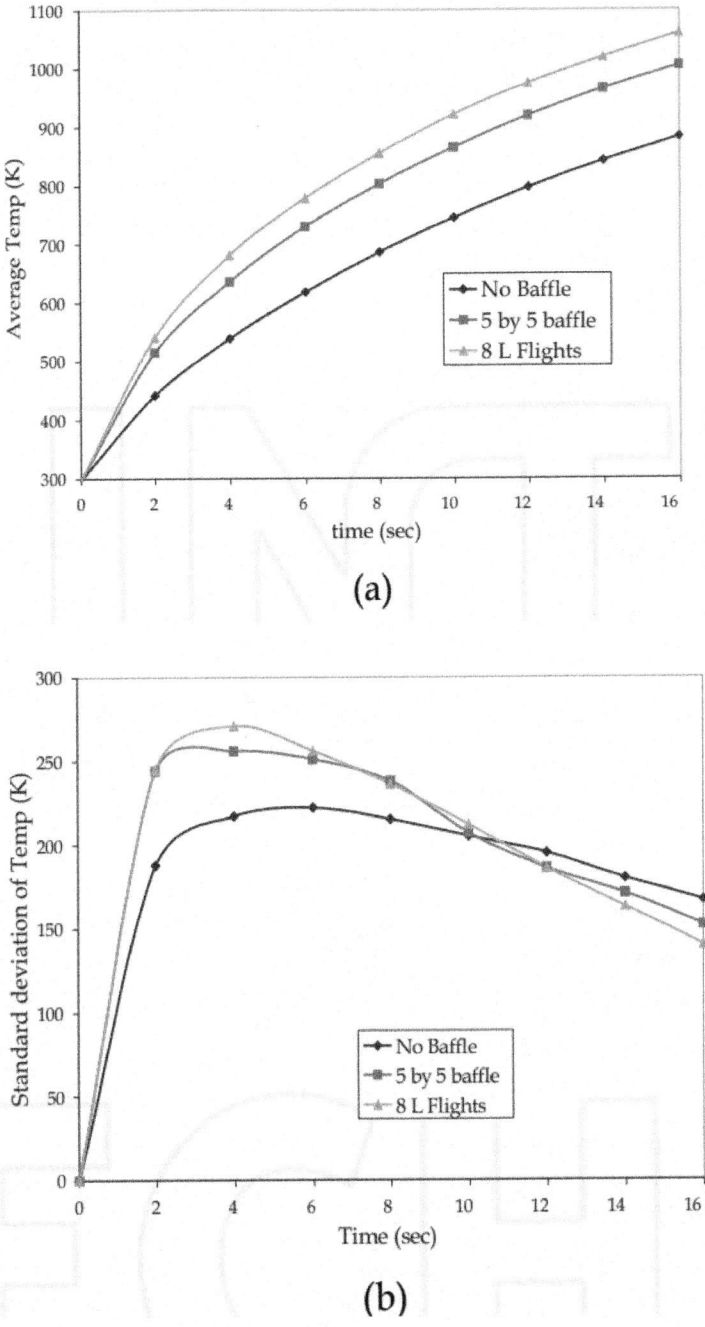

Figure 6: (a): Average temperature as function of time for different baffle configurations. (b): Standard deviation versus time for different baffle configurations.

Effect of baffle size on heat transfer in calciners

The effect of the size of the rectangular baffles/flights is investigated using DEM simulations. In Figure 7, the axial snapshots captured at time T= 0, 1 and 3 revolutions for 3 different baffle configurations: (i) no baffle (ii) 8 baffles of each 64 mm2 cross sectional area (iii) 8 baffles of each 400 mm2 cross sectional area. In our DEM model, four (2 by 2) and twenty-five (5 by 5) particles of radius 2 mm are glued together to form each of the baffles in case (ii) and (iii) respectively. The blue core signifies the mass of particles at initial temperature.

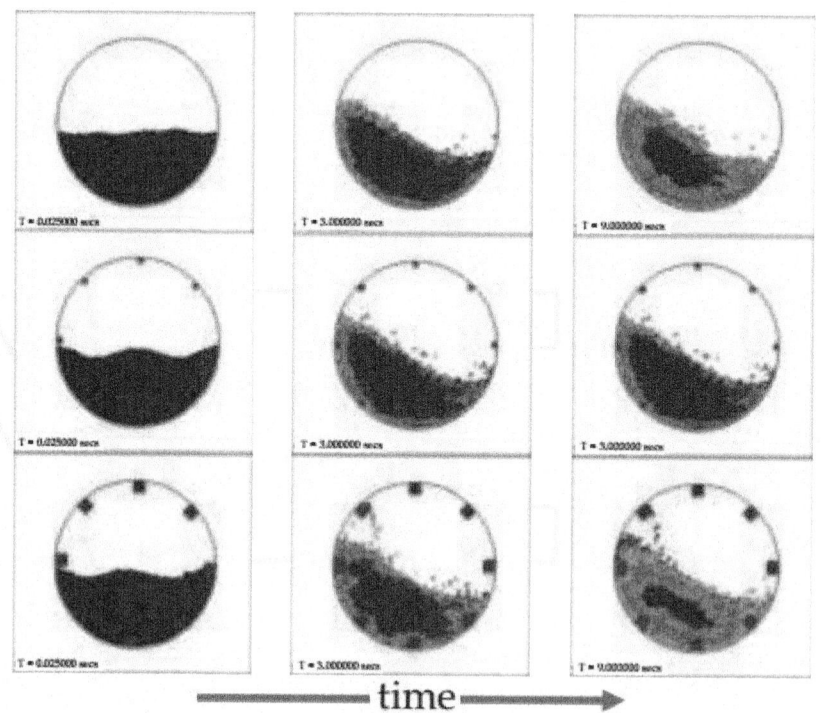

Figure 7: shows a time sequence of axial snapshots of color-coded particles in the calciner. Time increases from left to right (t = 0, 3 and 9 secs), while the baffle size increases top to bottom

This cold core shrinks with time for all cases, but it shrinks faster for a calciner with bigger baffles. The number of red particles in the bed also increases for calciners with baffles of bigger sizes. Thus, increased surface area of the bigger baffle enhances heat transfer within the calciners. The effect of baffle size on heat transfer is quantified by calculating the average bed temperature as a function of time for all baffle configurations. Average bed temperature rises faster for calciners with bigger baffles, as seen in Figure 8(a). The uniformity of the temperature of the particle bed is quantified by estimating the standard deviation of the bed temperature. Figure 8(b) shows the effect of the baffle size on the uniformity of the bed temperature, systems with bigger baffles reach uniformity quicker.

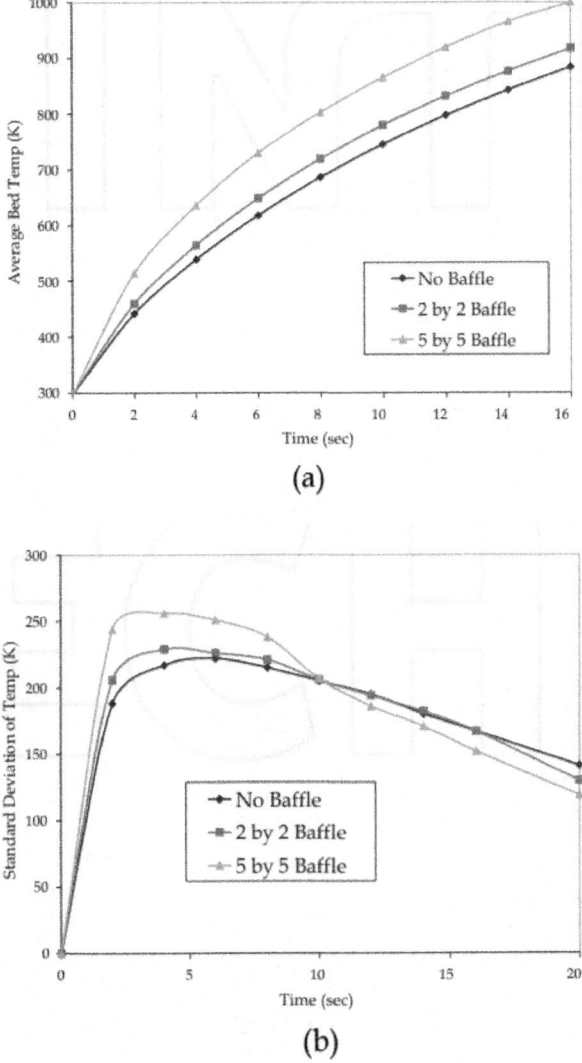

Figure 8: (a): Average bed temperature as a function of time for different sizes of rectangular baffles (b): The evolution of standard deviation versus time for different baffle configurations.

Effect of number of baffles/flights on heat transfer in calciners

The number of baffles is an important geometric parameter for the rotary calciner. The effect of the parameter has been investigated with a calciner with L-shaped baffles. We calculated the evolution of the temperature depicted in successive snapshots for 0, 4 and 12 baffles in Fig 9. There is a cold core which shrinks with time for all the cases, and it shrinks faster for the calciner with larger number of baffles. The number of red particles present in the bed also increases for calciners with more baffles. Thus, increase in the number of baffles causes enhancement in heat transfer within the calciners. The effect of number of baffles on heat transfer is quantified by calculating the average bed temperature as a function of time for all baffle configurations. The average temperature of the bed rises faster for calciners with higher number of baffles. This can be seen in Figure 10(a). The uniformity of the temperature of the powder bed is quantified by estimating the standard deviation of the surface temperature of the bed and it is shown in Figure 10(b). It can be seen that the thermal uniformity of the bed is directly proportional to the number of baffles.

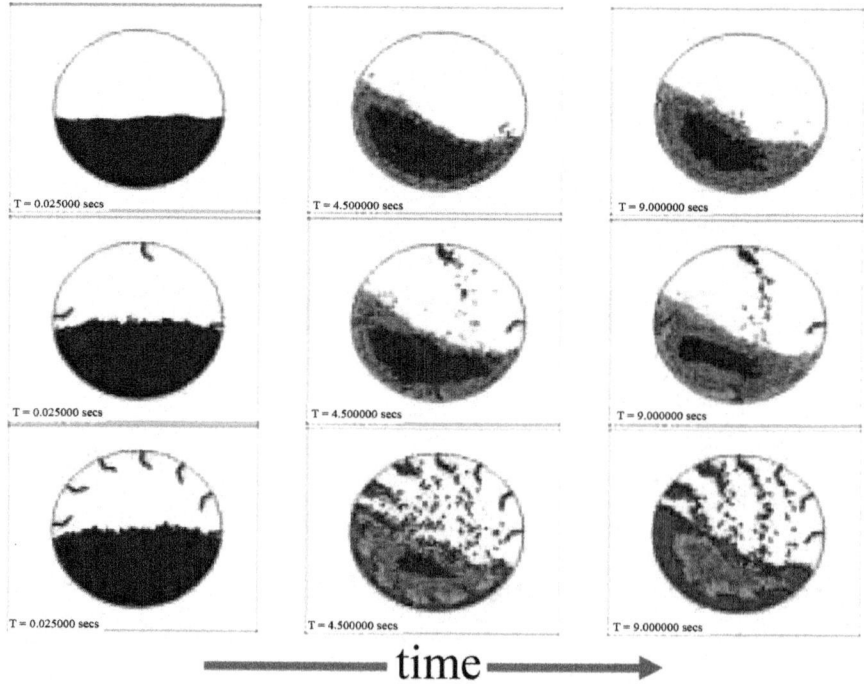

Figure 9: Evolution of the temperature for non-baffled and baffled calciners (with 4 and 8 flights) at time = 0, 1.5 and 3 revs.

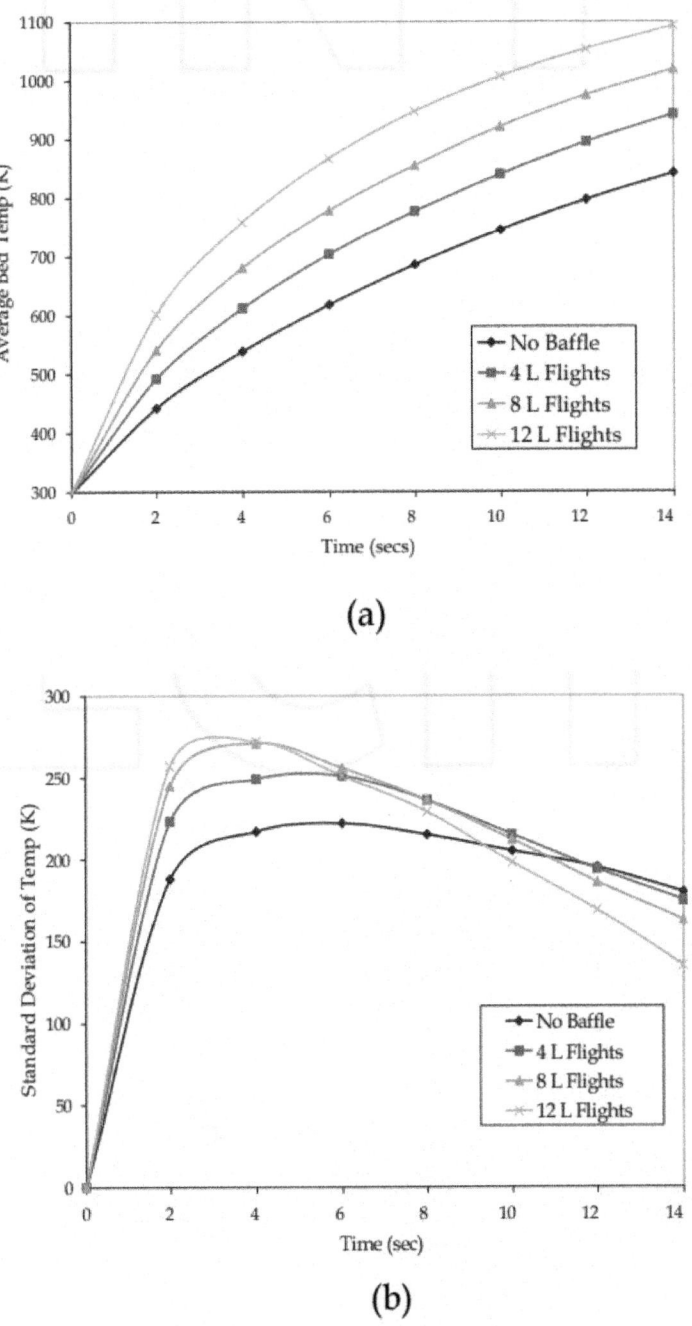

Figure 10: (a): Effect of number of L-Shaped baffles on heat transfer. (b): The evolution of standard deviation versus time for different baffle configurations

Effect of speed in baffled calciners

Heat transfer as a function of vessel speed is examined for L-shaped baffles. The evolution of temperatures of the particles is estimated for calciners with 8 flights/baffles rotated at different speeds: 10, 20 and 30 rpm (shown in Fig 11). The cold core gets smaller with time for all the cases, but this reduction is faster for calciners rotated at higher speed. The number of red particles present in the bed also increases for calciners rotating with higher speed. Thus, an increase in the speed enhances heat transfer within the calciners. The effect of speed on heat transfer is quantified by means of the average bed temperature as a function of time for all baffle configurations. Average bed temperature rises faster for calciners with higher speeds, as seen in Figure 12(a). This observation contradicts previous observations for un-baffled calciners. The higher vessel speed ensures more scooping of the material inside the bed and redistribution of the particles per unit of time, by the L-shaped baffles. The uniformity of the temperature of the particle bed is quantified by estimating the standard deviation of the surface temperature of the bed. As expected, bed rotated at higher speed reaches thermal uniformity faster (see Fig. 12(b)).

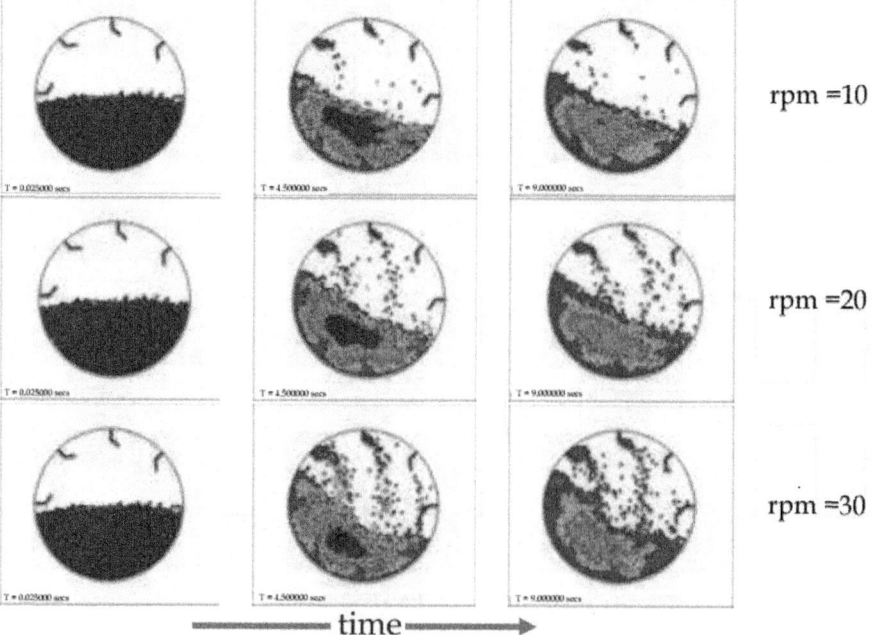

Figure 11: Evolution of temperature for baffled calciners (8 flights) at different rotational speeds of 10, 20 and 30 rpm, for different values of time = 0, 4.5 and 9 secs.

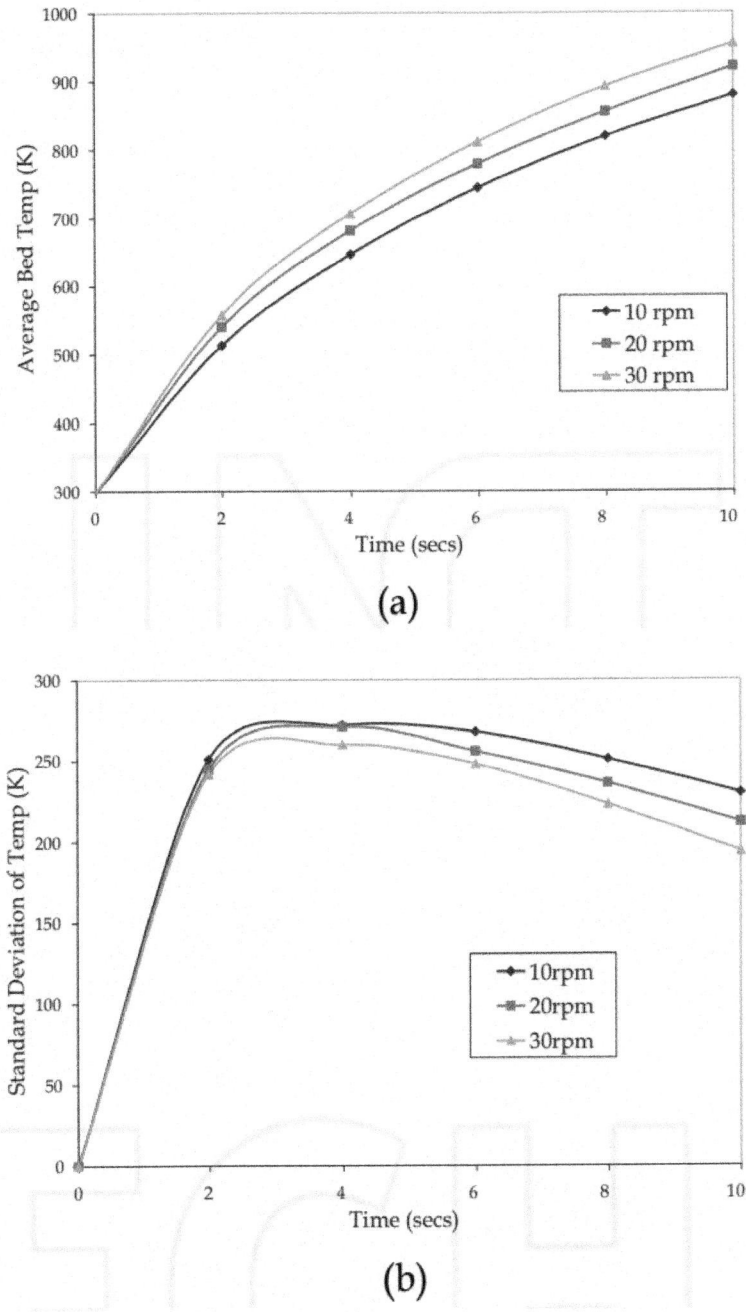

Figure 12: (a): Effect of speed on heat transfer for calciners with L-Shaped baffles. (b): The evolution of the standard deviation versus time for different vessel speeds.

Effect of adiabatic baffles on heat transfer in calciners

In the previous simulations all baffles were always at the wall temperature and enhanced the heat transfer and thermal uniformity (mixing) in the calciners. However, this can be due to two distinct effects. The flights not only scoop and redistribute particles enhancing convective transport, but also heat up the particles during the contact, increasing area for conductive transport. To nullify the conduction effect and check how flights affect convective heat transfer, L-shaped baffles were maintained at an adiabatic condition in a particle-baffle contact, $dQ = 0$ is considered. The 8 flights are thus maintained at the room temperature (298K) whereas, the wall remains at 1298 K. In Figure 13, the axial snapshots are displayed at time T= 0, 4 secs and 8 seconds for 2 different baffle configurations: (i) 8 Lshaped baffles at room temperature (298K) (ii) 8 L-shaped flights at the wall temperature (1298 K).

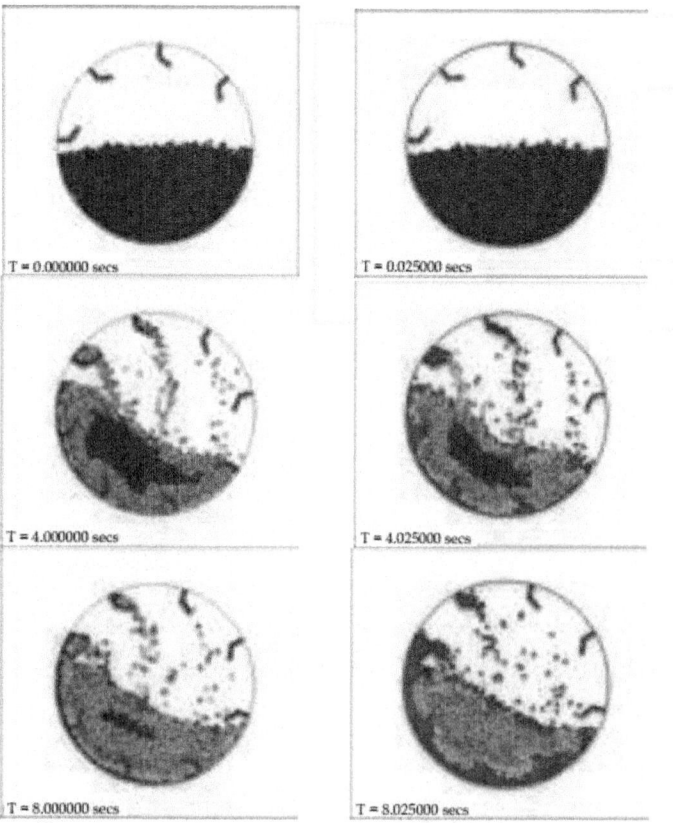

Figure 13: The axial snapshots of the calciners with cold (left) and hot baffles (right) at different time intervals.

The blue core signifies the mass of particles at the initial temperature. This cold core shrinks with time for all the cases, but it shrinks faster for calciners with L-shaped baffles at wall temperature. The number of red particles present in the bed also increases for calciners with L-shaped baffles at wall temperature (shown in the left column in Fig 13).

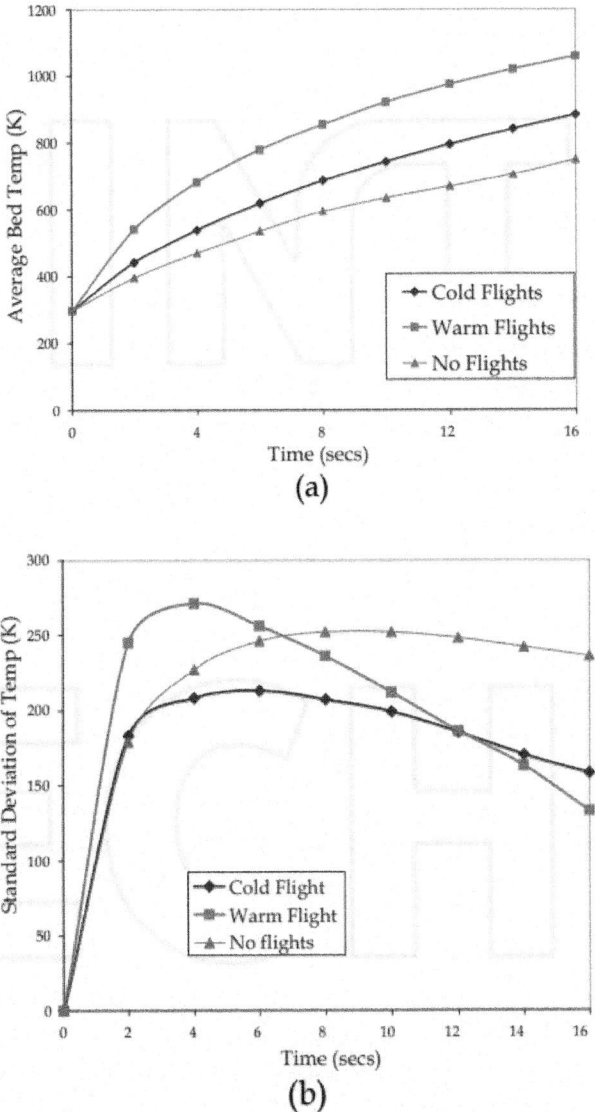

Figure 14: (a): The evolution of average bed temperature for 8 L shaped cold and warm flights and baffled calciners. (b): The evolution of thermal uniformity for calciners with cold, warm baffled and baffled flights.

Thus, heated baffles enhance heat transfer within the calciners. The effect of the temperature of the baffle on heat transfer is quantified by calculating the average bed temperature as a function of time for all baffle configurations and comparing it with the temperature profile of the non-baffled calciner. The average bed temperature rises faster for calciners with L-shaped baffles at wall temperature, but the calciner with colder baffle shows faster heat transport than non-baffled calciners (see Figure 14(a)). In Figure 14b, the uniformity of the temperature of the particle bed is presented by estimating the standard deviation of the surface temperature of the bed. The calciners with flights are reaching thermal uniformity faster than the non-baffled calciners. The temperature of the baffle does not cause much difference in thermal uniformity as both the curves for baffled calciners are very close to each other (convective mixing effect is independent of baffle temperature)

Heat transfer of copper particles in the calciner

Initially, 16,000 particles are loaded into the system in a non-overlapping fashion and allowed to reach mechanical equilibrium under gravitational settling. Subsequently, the vessel is rotated at given rate, and the evolution of the position and temperature of each particle in the system is recorded as a function of time. The curved wall is considered to be frictional. To minimize the finite size effects the flat end walls are considered frictionless and not participating in heat transfer. A parametric study was conducted by varying thermal conductivity, particle heat capacity, granular cohesion, vessel fill ratio, and vessel speed of the calciner. A cohesive granular material (K_{cohes} = 75, μ_{SP} = 0.8, μ_{DP} = 0.6, μ_{SW} = 0.8, μ_{DW} = 0.8) is considered to examine the effect of thermal properties and the speed of the vessel. Particles with temperature lower than 350°K are colored blue; those with temperature in between 350°K and 550°K are considered cyan. Those with temperature between 550°K and 750°K are considered green and for temperatures between 750°K and 950°K are considered yellow. Those particles with temperature higher than 950°K are colored red.

Effect of thermal conductivity

Three values of thermal conductivity of the solid material are considered: 96.25, 192.5, 385 W/m°K. The calciner is rotated at the speed of 20 rpm. As the heat source is the wall, the particle bed warms up from the region in contact to the wall. Particle-wall contacts cause the transport of heat from the wall to the particle bed. With subsequent particle-particle contacts, heat is transported inside the bed. In Figure 15a, the axial snapshots captured at time T= 0, 0.5 and 1 revolutions for varying thermal conductivities are displayed. The combination of heat transfer and convective particle motion results in rings or striations as the temperature decrease from the wall to the core of the bed. The presence of these concentric striations signifies that under the conditions examined here, the dominant mechanism is radial conductive transport of heat from the wall to the core of the bed. The blue core signifies the mass of particles at initial temperature. This cold core shrinks with time, as expected; the volume of the blue core shrinks faster for higher particle conductivities. The average bed temperature is illustrated in Figure (15d). As conductivity increases, the system exhibits faster heating. The variation of the standard deviation of the temperature of the bed is illustrated on Fig. 15 (e). Uniformity in the bed temperature increases with conductivity until the end of 5 revolutions. Finally the bed with higher conductivity rapidly reaches a thermal equilibrium with the isothermal wall, where all the particles in bed reach the wall temperature and there is no more heat transfer.

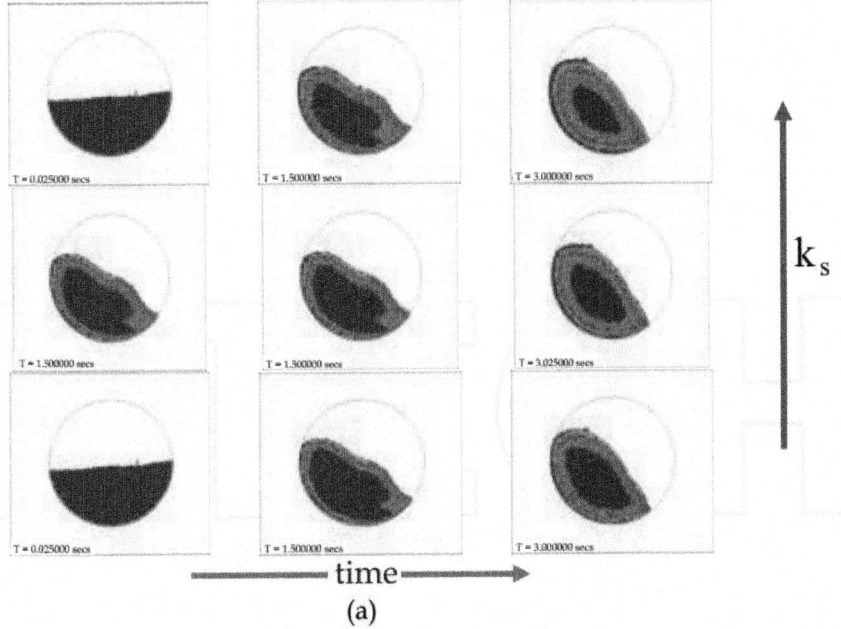

(a)

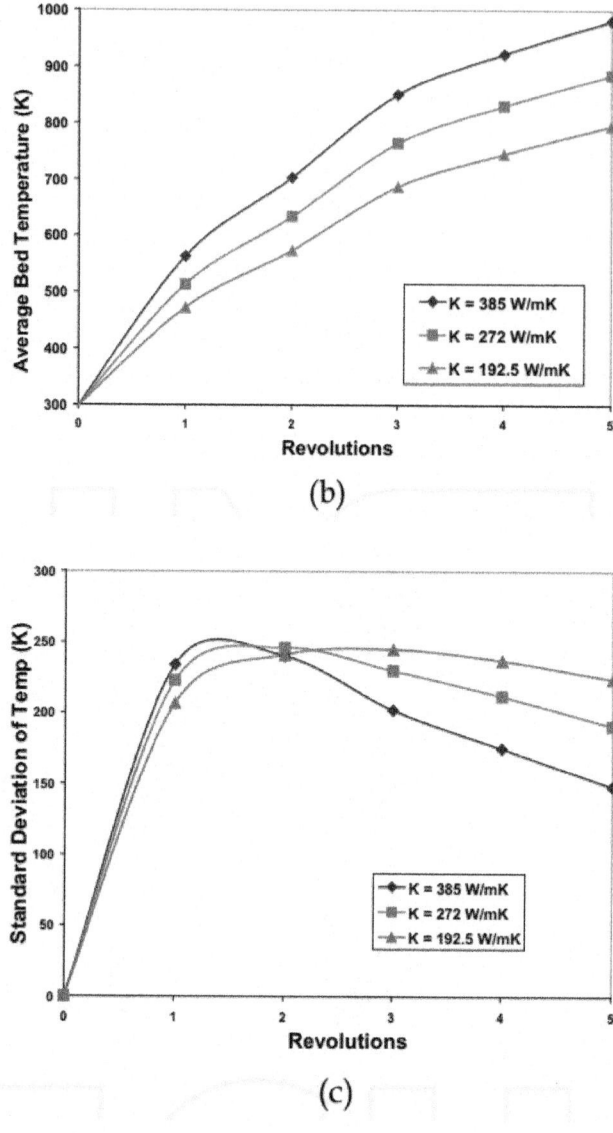

Figure 15: (a) shows a time sequence of axial snapshots of color-coded particles in the calciner. Time increases from left to right (t = 0, 1.5 and 3 secs), while the thermal conductivity increases from bottom to top (ks = 192.5, 272, 385 W/mK). (b) shows the growth of average bed temperature over time for materials with different conductivity. The granular bed heats up faster for material with higher conductivity. (c). illustrates the variation of the standard deviation of particle temperature over time for different conductivities. More uniformity of temperature in the bed for material of higher thermal conductivity.

A physical formula to fit the simulation prediction is derived based on the Leven berg Marquardt method, which uses non-linear least square based regression techniques. This curve fitting method is employed for the average bed temperature data displayed in Fig 1 for the highest thermal conductivity (ks = 385 W/mK). The 3rd order polynomial derived is as follows

$$T_{avg} = 301.92 + 288.624n - 45.05n^2 + 2.9n^3$$

(6)

where T_{avg} and n are the average bed temperature and number of revolutions respectively. The vessel speed for this data is 20 rpm and so n=1 corresponds to 3 seconds. The correlation coefficient for this fit R = 0.9989. The simulation data and the 3rd order least square fit curve of the data are illustrated in Fig. 15d.To gather an insight of the evolution of average bed temperatures beyond 5 revolutions, the average bed temperatures at all-time intervals for each of the cases in Fig. 15b is scaled by the corresponding average temperature at 5 revolutions. In Fig. 15e, almost all of the data points for different conductivity overlap showing the evolution of average temperature follow the same shape and will reach thermal equilibrium with the wall at the same rate shown in Fig. 15b.

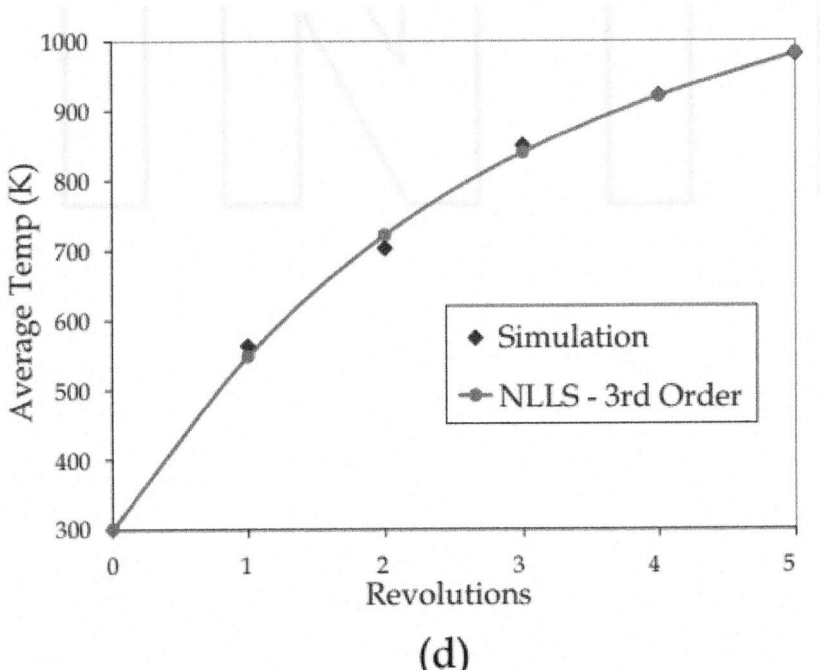

(d)

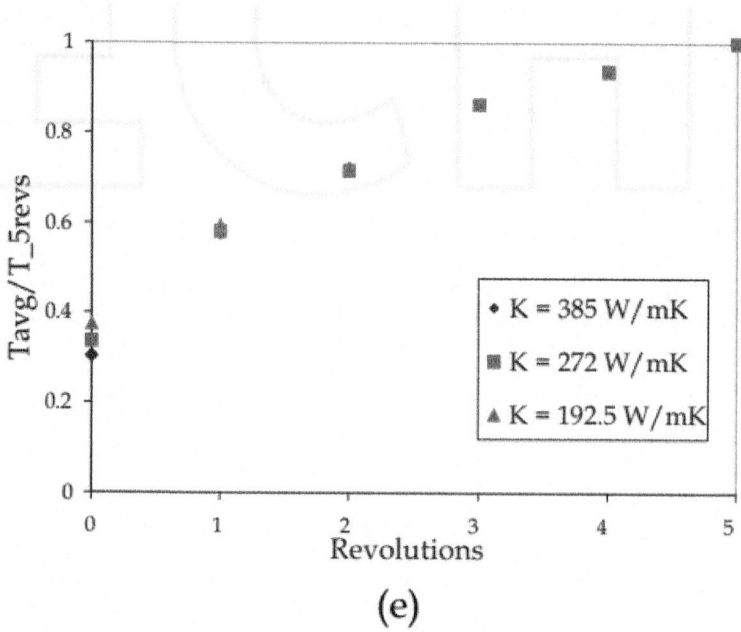

(e)

Figure 15: (d): Comparison of the simulation data (for ks = 385 W/mK) and the non-linear least square fit, (e): Average bed temperature over time for materials with different conductivities.

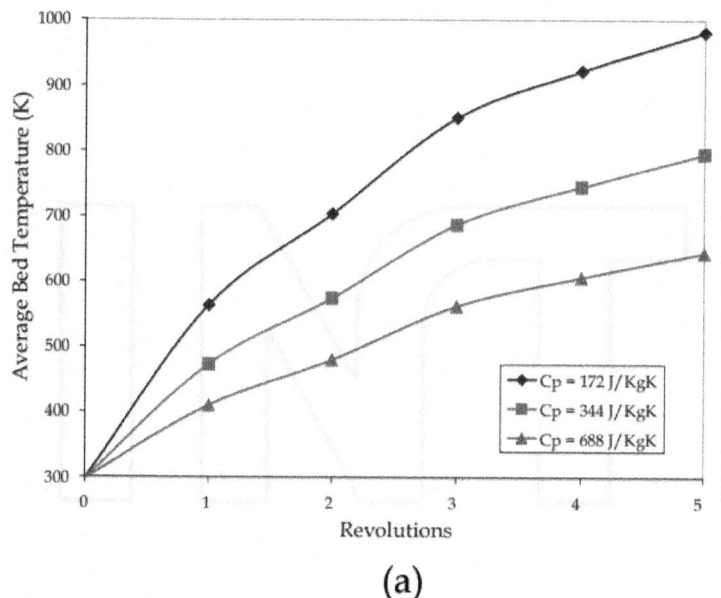

(a)

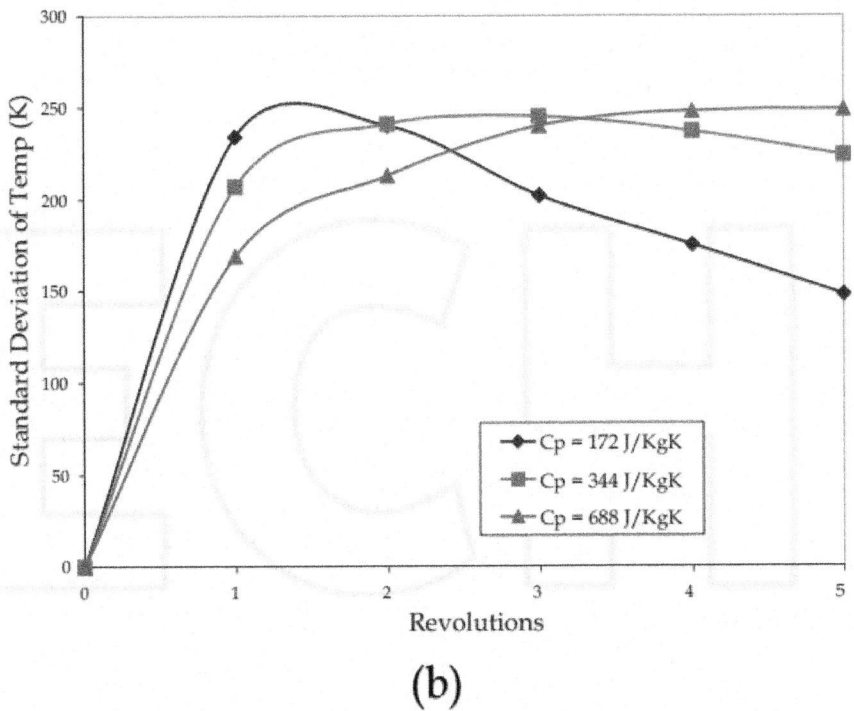

(b)

Figure 16: (a) Evolution of average bed temperature over time in a calciner, for material with different heat capacities (Cp= 172, 344, 688 J/KgK). Granular bed heats up faster for material with lower heat capacity (b) Variation of the standard deviation of particle temperature over time for different heat capacities. More uniformity of temperature is seen in the bed for materials of lower heat capacity.

Effect of heat capacity

After quantifying the effect of thermal conductivity, the other main thermal property of a material, heat capacity, is checked. Three values of heat capacity of the granular material are considered: 172, 344 and 688 J/Kg°K, while keeping the thermal conductivity constant at 385 W/moK. Once again, the calciner is rotated at the speed of 20 rpm. Average bed temperatures are estimated as a function of time (Fig. (16a)). As expected, particles with lower heat capacity exhibit faster heating. The evolution of the standard deviation of temperature of the granular bed is illustrated in Figure 16(b). The variability in the bed temperature is larger for the material with lower heat capacity until 2 revolutions, but at the end of 5 revolutions, more uniform temperature is observed for the material of lower heat capacity

Effect of granular cohesion and friction

The effect of granular cohesion on heat transfer is examined while keeping the thermal properties constant (k_s = 385 W/m°K and C_p = 172 J/Kg°K). As discussed in Section 2, to simulate different levels of cohesion and friction, the bond number K, the coefficients of static and dynamic friction between particles (μ_{SP} and μ_{DP}) and the coefficients of static and dynamic friction between particle and wall (μ_{SW} and μ_{DW}) are varied. Heat transfer in cohesionless particles (K$_{cohes}$ = 0, μ_{SP} = 0.8, μ_{DP} = 0.1, μ_{SW} = 0.5, μ_{DW} = 0.5) is compared with a slightly cohesive (Kcohes = 45, μ_{SP} = 0.8, μ_{DP} = 0.1, μSW = 0.5, μ_{DW} = 0.5) and a very cohesive material (K$_{cohes}$ = 75, μ_{SP} = 0.8, μ_{DP} = 0.6, μ_{SW} = 0.8, μ_{DW} = 0.8). The evolution of the average bed temperature over time is shown in Fig. 17a. For all cases examined here, cohesion does not cause a significant difference in the temperature profiles. The variability in bed temperature is quantified by the standard deviation of the particle temperature. In Figure 17b, the variation in standard deviation of temperature for the three values of cohesion is shown. Once again for the cases examined here, granular cohesion does not have a significant effect in the uniformity of the particle temperature of the bed.

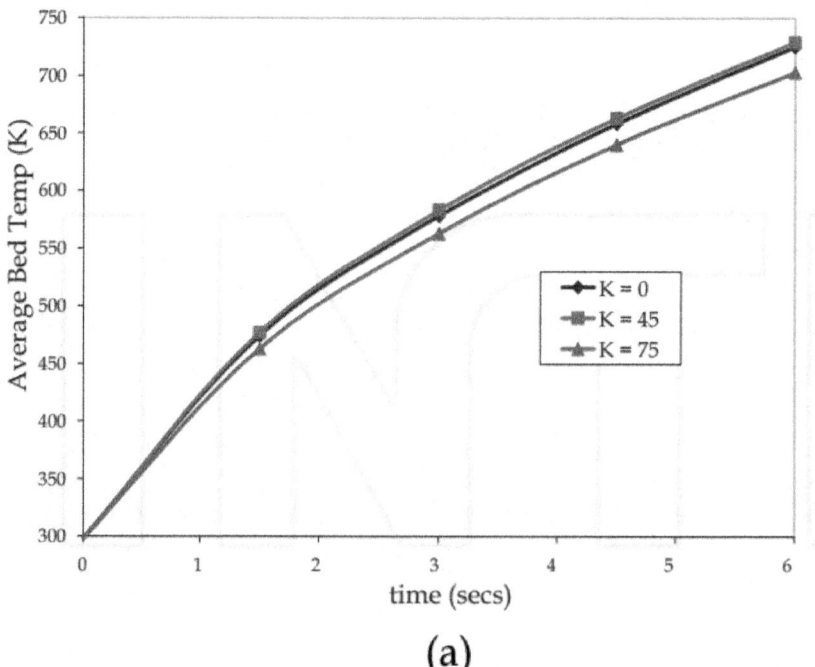

(a)

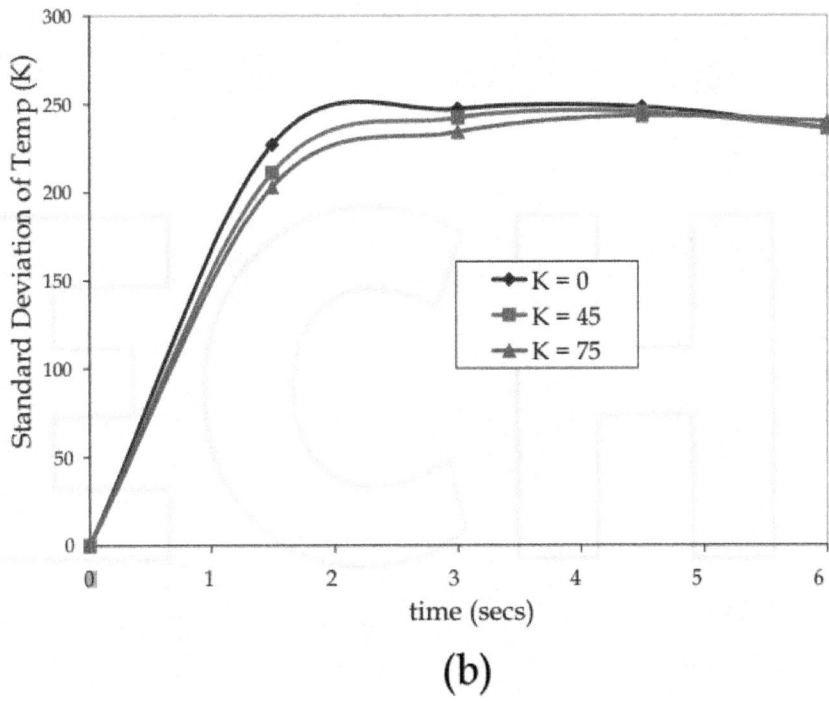

(b)

Figure 17: (a) shows the evolution of average bed temperature over time in the calciner, for materials with different granular cohesion (K_{cohes} = 0, 45, 75). (b) illustrates the variation of the standard deviation of particle temperature over time for different levels of granular cohesion. Granular cohesion has no significant effect in heat transfer.

Effect of vessel speed

In order to examine the effect of vessel speed, the most cohesive granular system (K_{cohes} = 75) is rotated at three different speeds: 12.5, 20 and 30 rpm, for thermal transport properties constant and equal to: k_s = 385 W/moK and C_p = 172 J/KgoK. Figure 18a displays snapshots captured at 0, 0.5 and 1 revolutions for varying vessel speeds. The higher vessel speed applies a higher shear rate to the granular system, causing significant differences in flow behavior, evident in the different dynamic angle of repose of the bed at each rotational speed.

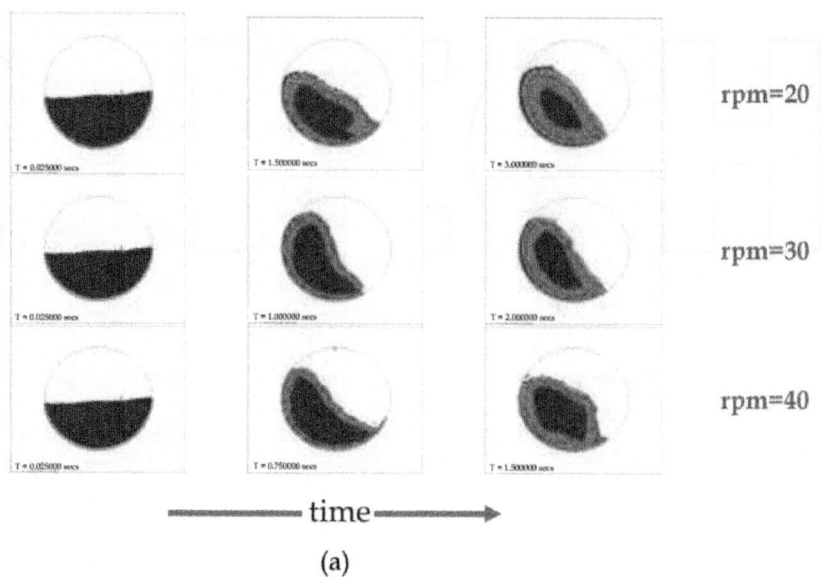

(a)

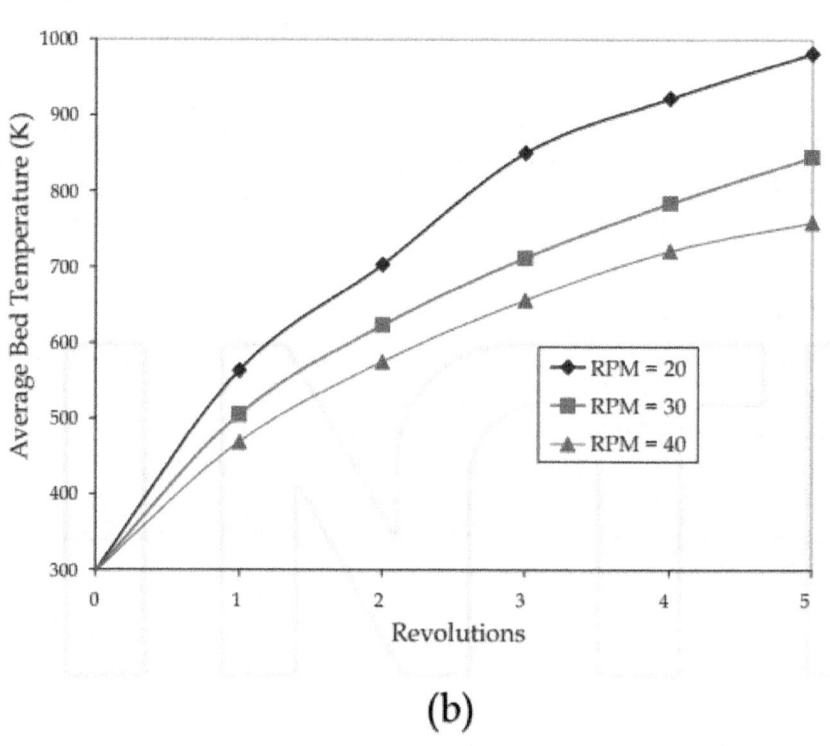

(b)

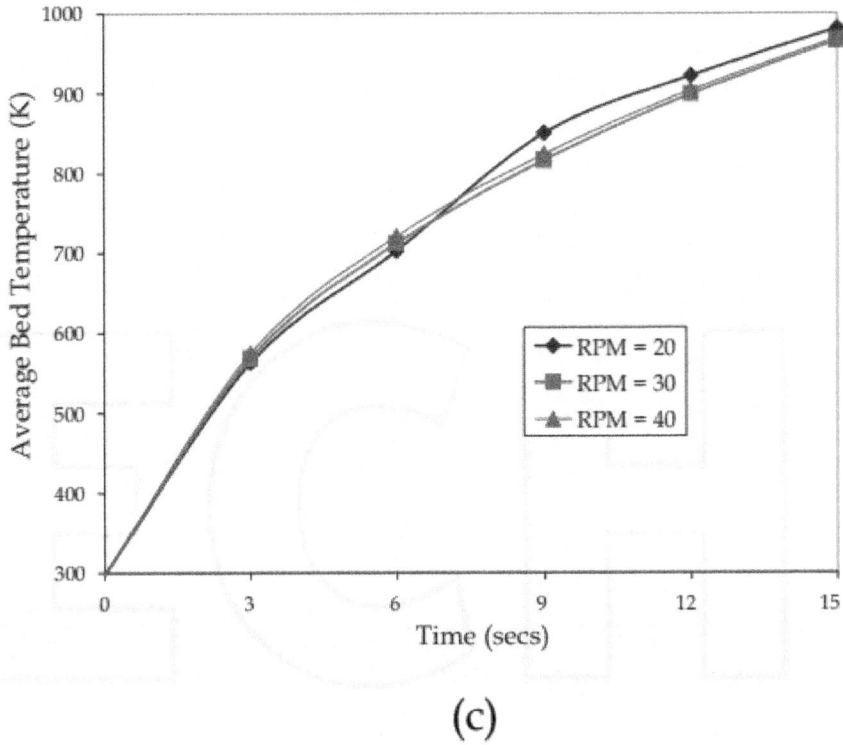

(c)

Figure 18: (a) shows the time sequence of axial snapshots of color-coded particles in the calciner. Time increase from left to right hand side (T = 0, 0.5 and 1 revolution), while the vessel speed increases from top to bottom (20, 30 and 40 rpm). (b) shows the evolution of average bed temperature versus vessel rotations for different vessel speeds. (c) shows the average bed temperature profile over real time for different vessel speeds. Rotation speed increases heat transfer in a per-revolution basis but the effect disappears on a per-time basis.

On a per-revolution basis, slower speed caused higher temperature rise (as shown in Fig. 18(b)). A thicker red band of particles (adjacent to the wall) and a smaller blue core are evident. At slower speeds, each particle has a more prolonged contact with the heated wall, which contributes to the rapid rise in the temperature. However, when analyzed on per absolute time basis, the effect of speed dissappers as the average bed temperatures for all the cases follows nearly identical trends (Fig 18(c)). The standard deviation of the temperature of the bed is also estimated in per-revolution and per-time basis. While the temperature of the bed is more uniform at slower speeds on a per revolution basis (Fig. 19(a)). This effect almost disappears on the real time basis (Fig 19(b)).

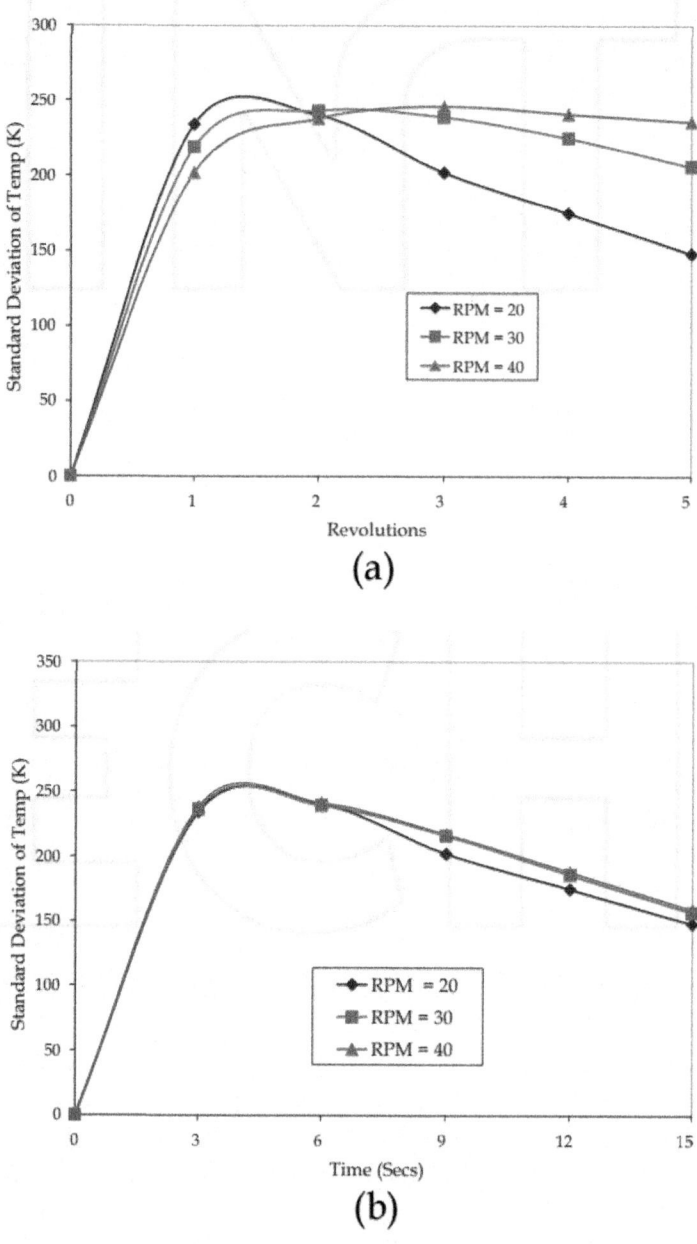

Figure 19: (a) shows the evolution of standard deviation of bed temperature versus vessel rotations for different vessel speeds. (b) Shows the standard deviation of bed temperature over real time for different vessel speeds. Rotation speed increases the uniformity of bed temperature in a per revolution basis but the effect almost disappears on a per-time basis.

Effect of fill ratio

Three different fill levels, 18%, 43% and 56%, are simulated using 7000, 16000 and 20,000 particles. Once again, the vessel is rotated at 20 rpm. Particle's thermal transport properties remain constant at k_s = 385 W/m°K and Cp = 172 J/Kg°K. Non-cohesive conditions are considered. In Fig. 20(a), the change in average bed temperature with time is shown as a function of the fill ratio. As expected, the granular bed with lower fill fraction heats up faster. Faster mixing is achieved for the lower fill fraction case, which causes rapid heat transfer from the vessel wall to the granular bed. The temperature is more uniform for lower fill fraction at the end of 5 revolutions (Fig 20(b)).

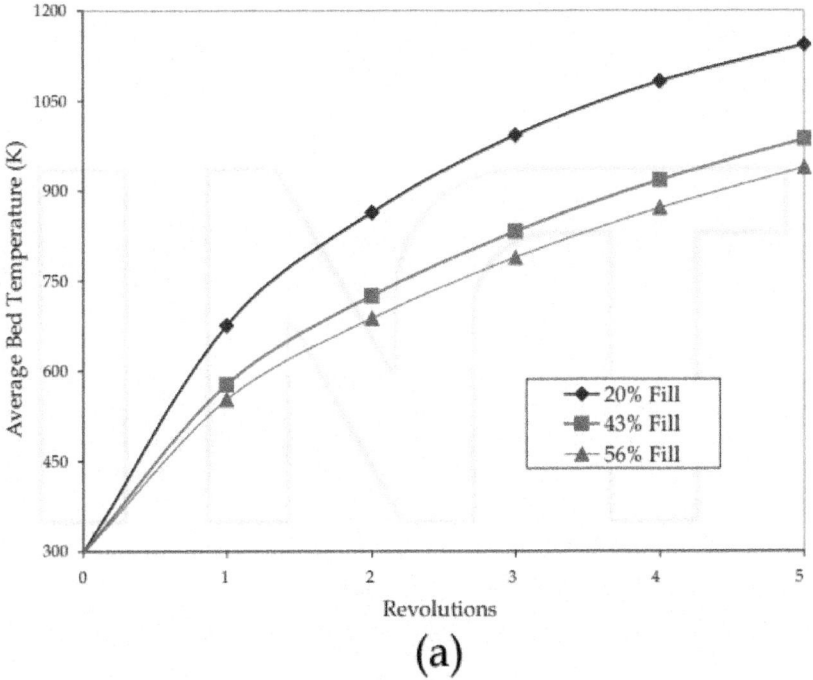

(a)

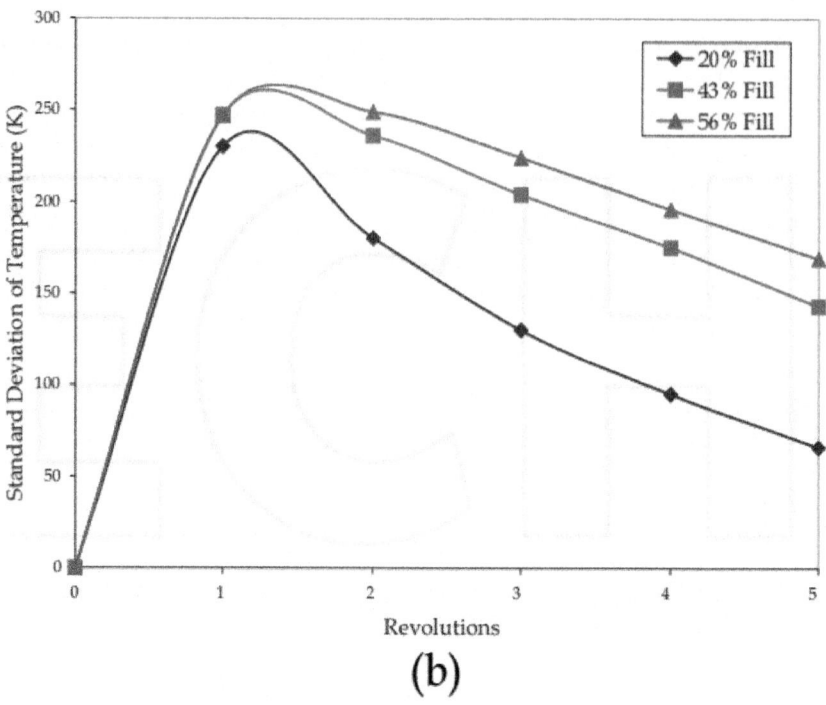

(b)

Figure 20: (a) shows the evolution of average bed temperature over time for granular bed of different volumes (fill % = 20, 43, 56). The granular bed heats up faster for lower fill fraction. (b) illustrates the variation of the standard deviation of particle temperature over time for different fill fractions. More uniformity of temperature in the bed of lower fill fraction.

Heat transfer in a double cone impregnator

We simulate the flow and heat transfer of 18,000 particles of 3mm size rotated in a double cone impregnator equipped with a baffle of variable size. Initially, particles are loaded into the system (with and without baffles) and allowed to reach mechanical equilibrium. Subsequently, the temperature of the vessel (and the baffle) is raised to the desired value of 1298°K, and the evolution of the temperature of each particle in the system is recorded as a function of time. All impregnator walls are considered to be frictional in the simulation. Coefficients of static friction between particles and particle-wall are considered to be 0.8 and 0.5 respectively. Coefficient of dynamic friction is considered to be the same as those of static friction for simplicity.

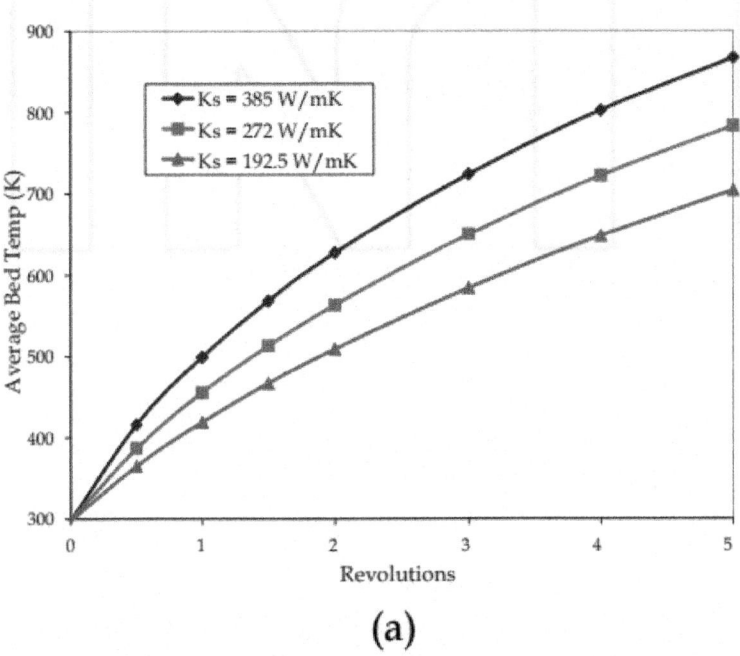

(a)

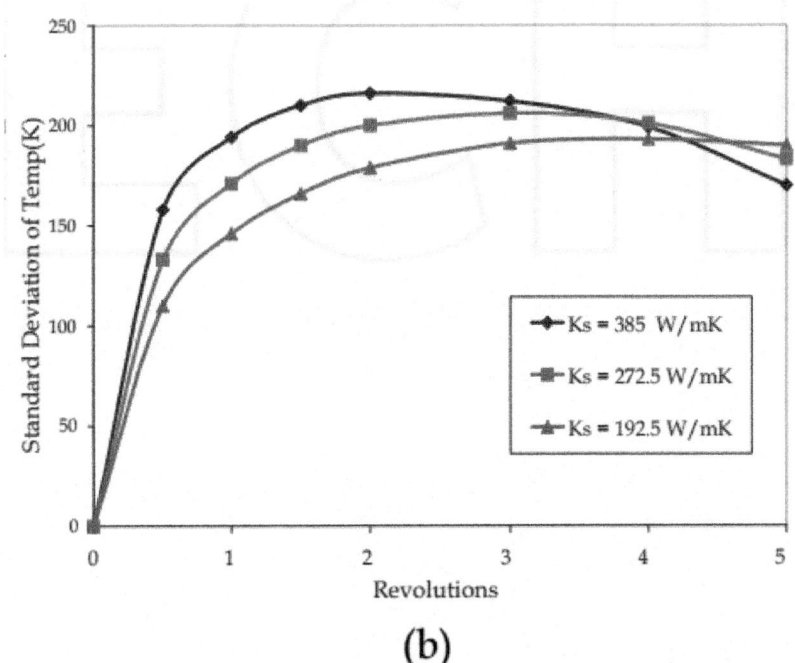

(b)

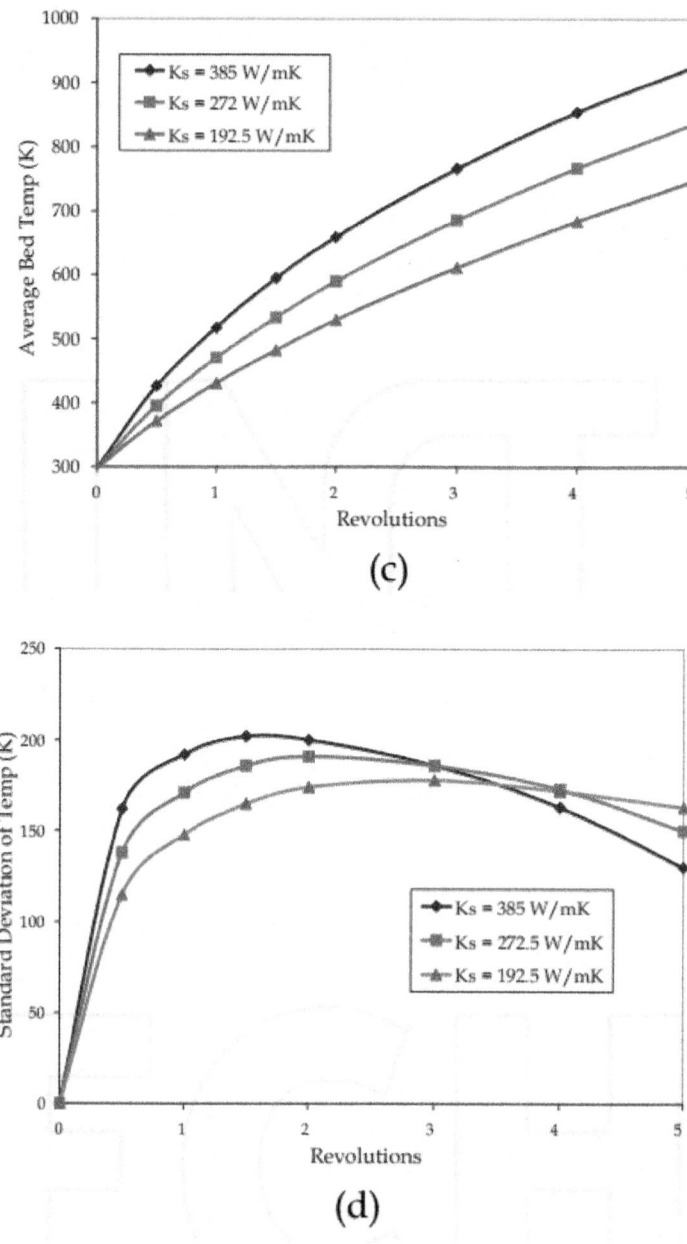

(c)

(d)

Figure 21: (a) shows the growth of the average bed temperature over time in a non-baffled impregnator, for materials with different thermal conductivities (k_s = 192.5, 272, 385 W/mK). (b) shows the growth of the average temperature of the bed over time in baffled impregnator, for materials with different thermal conductivities (k_s = 192.5,

272, 385 W/mK). (c) illustrates the variation of the standard deviation of particle temperature in the no baffled impregnator over time for different thermal conductivities. (d) illustrates the variation of the standard deviation of particle temperature over time for different thermal conductivities for a baffled vessel. Granular bed heats up faster for material with higher thermal conductivity.

Firstly, the effects of thermal conductivity and heat capacity on temperature are examined. Subsequently, the impact of the vessel speeds and baffle size on heat transfer rate and temperature field uniformity are examined. Three cases are considered: (a) no baffle, (b) baffle with 9 cm^2 cross-section (c) baffle with 36 cm2 cross-section. Particles in all the impregnator simulations are considered non-cohesive. Once again, the initial temperature of all the particles is considered to be at 298°K (room temperature) whereas the temperature of the wall (and the baffle if present) is considered to be at 1298°K (and in isothermal condition). Particles with temperature lower than 400°K are colored blue, while those with temperature in between 400°K and 600°K are colored green; those with temperature in between 600°K and 900°K are colored yellow, and those with temperatures higher than 900°K are colored red.

Effect of thermal conductivity

Higher thermal conductivity favored the transfer of heat and enhanced temperature uniformity in calciner flows. Impregnators and calciners both tumble but have different shapes. The effect of thermal conductivity on granular bed temperature is quantified for non-baffled and baffled impregnators. Three values of thermal conductivity (ks) of the material are considered: 96.25, 192.5 and 385 W/m°K. All three simulations are performed at 20 rpm. The evolution of the average bed temperature as a function of thermal conductivity is shown in Fig. 21a (un-baffled impregnator) and 21b (baffled impregnator with baffle cross-sectional area of 9 cm2). More conductive particles exhibit faster heating in both cases. The standard deviation plots corresponding to particle temperature for the non-baffled and baffled impregnators are shown in Fig 21c and 21d. In the first three revolutions, we observe more uniform (lower standard deviation) temperature for the cases with lower conductivity. At later times, as most particles reach high temperatures, all the curves show low values of standard deviation (not shown). More uniform temperature is attained in the bed of highest conductivity (ks = 385 W/mK) after 3 revolutions (Figs. 21c and 21d).

Effect of heat capacity

Similar to the transient heat transfer analysis of the calciner, the effect of heat capacity is also quantified for non-baffled and baffled impregnators.

Three values of heat capacity are considered: 172, 344 and 688 J/Kg°K. The coefficient of thermal conductivity is kept constant at 385 W/m°K. These simulations are performed at a vessel speed of 20 rpm. The evolution of the average bed temperature over time is depicted in Fig. 22(a). As expected, the lower the heat capacity, the faster the rise in bed temperature. The variability of the bed temperature is higher for material with higher heat capacity (see Fig. 22(b)).

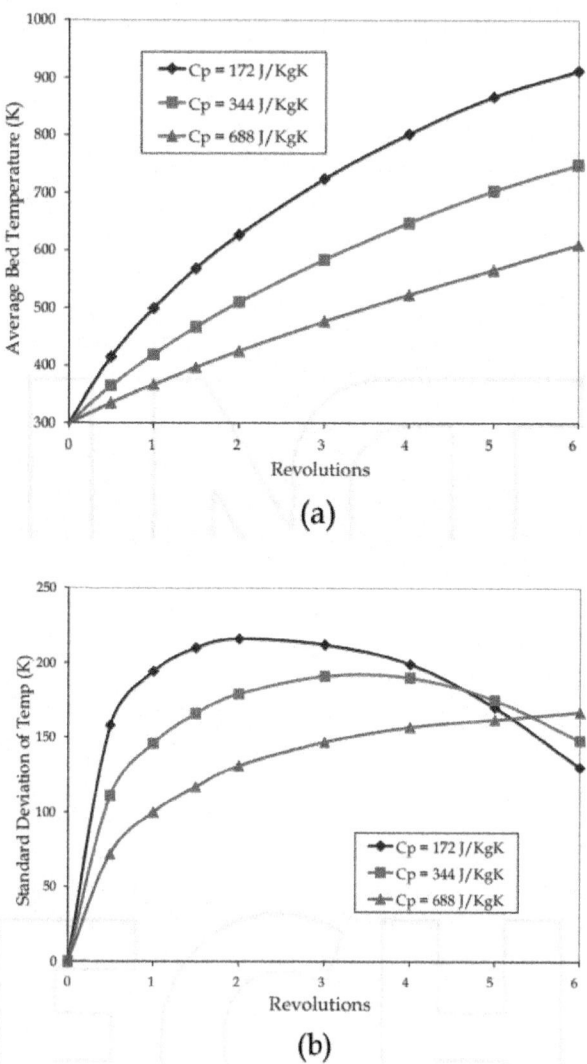

Figure 22: (a) shows the growth of average bed temperature over time for materials with different heat capacities (Cp =172,344,688 J/KgK). (b) illustrates the variation

of the standard deviation of particle temperature over time for different baffle sizes. Granular bed heats up faster for material of lower heat capacity.

Effect of baffle size

The baffle size is a significant geometric parameter for the impregnator. The effects of baffle size on the motion and heat transfer of the granular bed are studied for three cases: (a) no baffle (b) baffle with 9 cm^2 cross-section (c) baffle with 36 cm2 cross-section. Simulations are carried out for a vessel speed of 20 rpm. Thermal transport properties of the particles are ks = 385 W/m°K and C$_p$ = 172 J/KgoK. Figure 23(a) displays snapshots captured at 0, 2.5 and 4 seconds for varying baffle sizes. The effect of the baffle is clearly seen at the snapshot corresponding to 2.5 seconds. Red particles (T > 900°K) group around the baffle boundary, whereas the un-baffled container barely contains any red particles. The average bed temperature and the standard deviation of the temperature are shown in Figs 23(b) and 23(c), respectively. The temperature increases at a faster rate (Fig. 23(b)) and more uniformly (Fig. 23(c)) for cases with baffles. The temperature rise is faster for larger baffles. Baffles enhance mixing, which increase the uniformity in the temperature field of the granular bed. This observation is line with the experimental finding by Brone, 2000 and the numerical mixing results of Muguruma, 1997, who found that an optimal baffle size and positioning enhances mixing in rotary vessels.

Effect of vessel seed

Numerical simulations of heat transfer in calciners (shown above), reveal that lower speed enhances heat transfer and temperature uniformity on a per-revolution basis. Heat transfer as a function of vessel speed is studied here for a baffled (36 cm2 cross sectional area) impregnator at three different rotational speeds: 12.5, 20 and 30 rpm. Transport properties are: k$_s$ = 385 W/moK and C$_p$ = 172 J/Kg°K, respectively. Figure 24(a) displays snapshots captured at 0, 0.5 and 1 revolutions for varying vessel speeds. At the end of 0.5 and 1 revolution, more red particles (T > 900oK) are seen for vessels rotating at lower speeds; once again, on a per-revolution basis, lower speed causes higher temperature rise (as shown in Fig. 24(b)). Again, at slower speeds, the particles have a more prolonged contact with the heated wall, which contributes to the rapid rise in the temperature. However, on a per-time basis, the effect of speed partially disappears, as the average bed temperatures rise in unison for first 3 sec (Fig 24(c)) at all rotation speeds. However, after 3 seconds, temperature again rises faster for the vessel rotated at the slower speed of 12.5 rpm giving a counter-intuitive result. The standard deviation of the temperature of the granular bed is also estimated in per-revolution and per-time basis. The uniformity of the temperature of the bed significantly increases with slower speed of the calciner on a per-revolution basis (Fig. 25(a)). This effect also remains prominent on a real-time basis (Fig 25(b)).

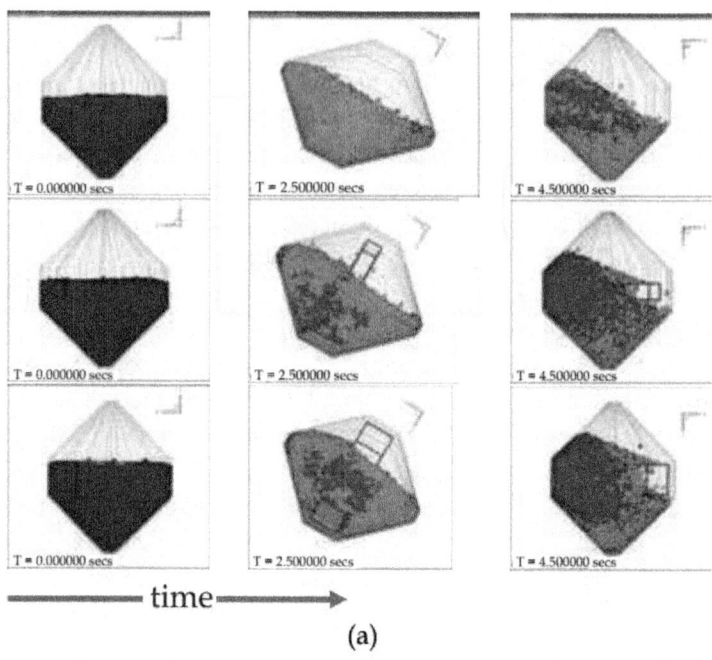

time ⟶

(a)

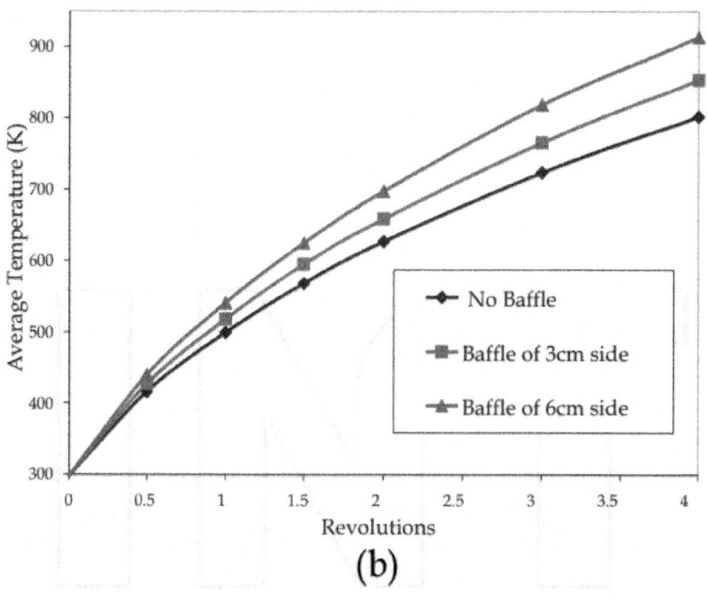

(b)

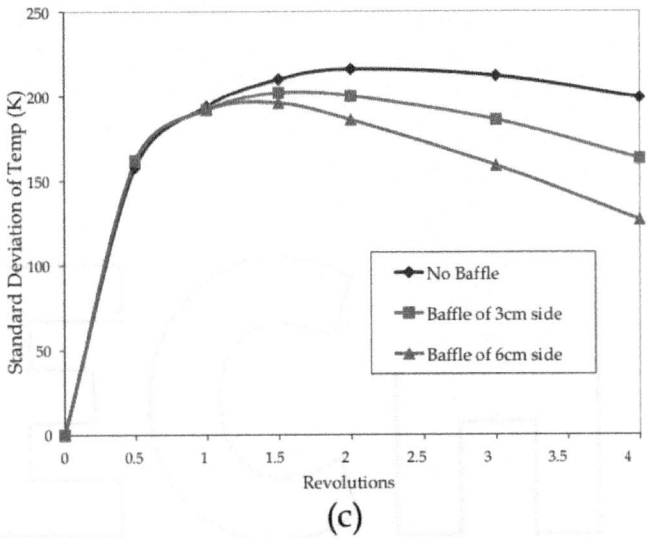

(c)

Figure 23: (a) shows a time sequence of simulation snapshots of color-coded particles in the impregnator. Time increase from left to right hand side (T = 0, 2.5 and 4.5 secs), while the baffle size increases from top to bottom (no baffle, baffle of 3cm , baffle of 6cm). (b) shows the growth of average bed temperature over time for different sizes of baffle. Depending on size, baffles can either increase or decrease heat transfer. (c) shows the variation of the standard deviation of particle temperature over time for different baffle sizes. Uniformity in the bed temperature is more with bigger baffle.

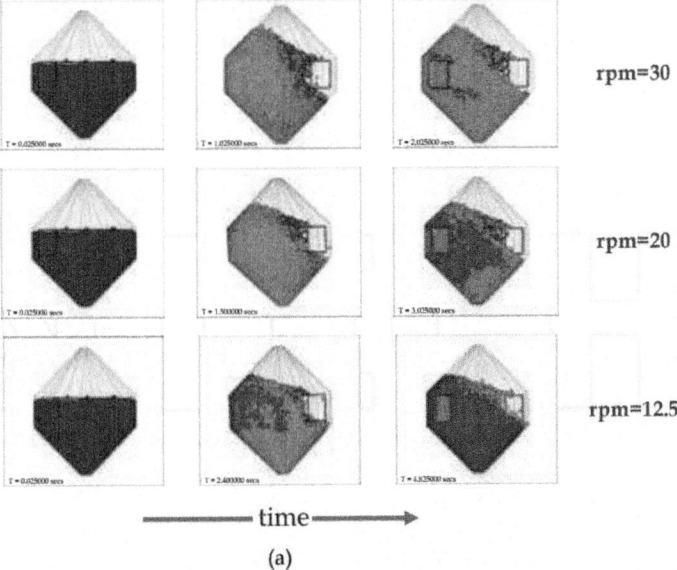

(a)

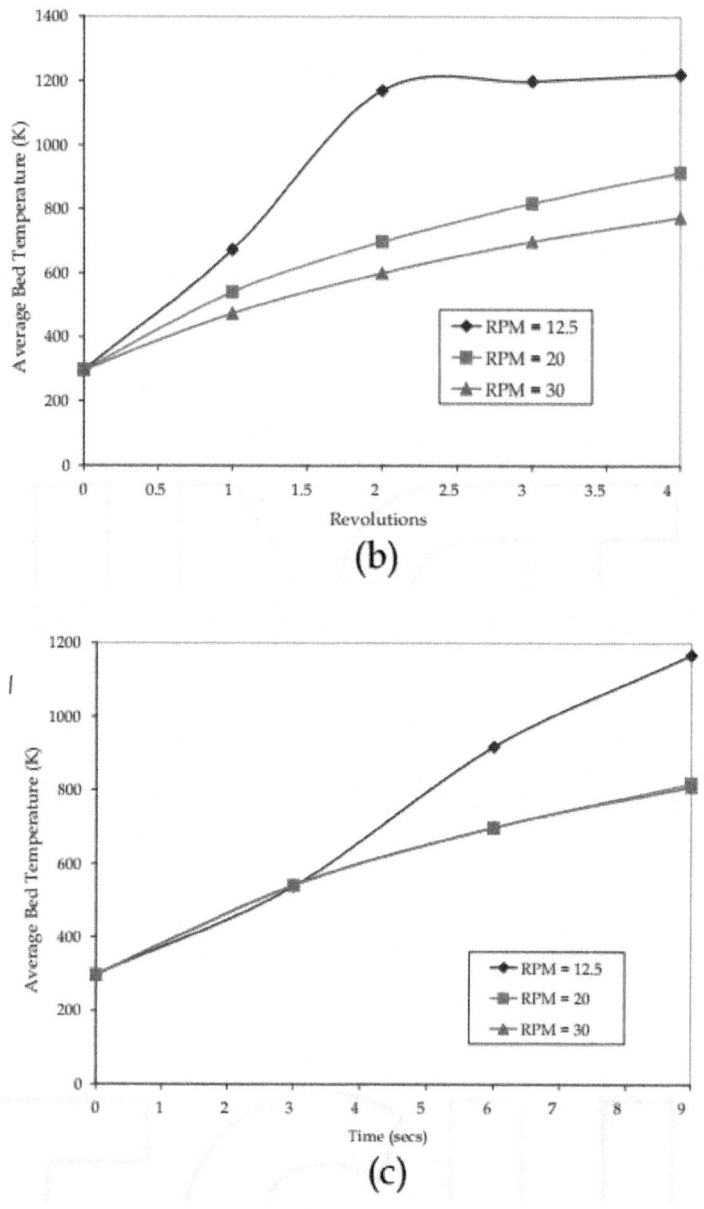

Figure 24: (a) shows a time sequence of simulation snapshots of color coded particles in the impregnator. Time increases from left to right hand side (T = 0, 0.5 and 1 revolution), while the vessel speed increases from bottom to top (12.5, 20, 30 rpm). (b) shows the evolution of the average bed temperature versus the number of revolutions for different vessel speeds. (c) shows the growth of average bed temperature over real time for different vessel speeds. Lower rotational speed of the impregnators facilitates

heat transfer if analyzed with respect to each revolution, but the effect disappears on a per-time basis.

DIMENSIONLESS ANALYSIS

Although conductivity and heat capacity of the material have similar effects in the heat transfer for both the calciner and impregnator, the vessel speed only affects the heat transfer in impregnators. Thus for impregnators we could define a parameter Π, which is directly proportional to the thermal conductivity (k_s) but inversely proportional to the specific heat (C_p) and the rotational speed (N) of the vessel. The heat balance of the granular material could be expressed as:

$$\rho C_p \left[\frac{\delta T}{\delta t} + v \bullet \nabla T \right] = k_s \nabla^2 T$$

or

$$\rho C_p \frac{T^0}{\tau} \frac{\delta T^*}{\delta t^*} + \rho C_p \frac{UT^0}{L} v^* \bullet \nabla^* T^* = k_s \frac{T^0}{L^2} \nabla^{*2} T^*$$

or

$$\frac{\delta T^*}{\delta t^*} + \frac{U\tau}{L} v^* \bullet \nabla^* T^* = \frac{k_s \tau}{\rho C_p L^2} \nabla^{*2} T^*$$

T^0, U and L are the initial bed temperature, linear velocity and the length of the vessel respectively. If we consider $\tau = L\,U$ (no independent time scale, i.e., Strouhal number $= 1$), the equation above becomes:

$$\frac{\delta T^*}{\delta t^*} + v^* \bullet \nabla^* T^* = \frac{k_s}{\rho C_p L U} \nabla^{*2} T^*$$

or

$$\frac{D T^*}{D t^*} = \Pi \nabla^{*2} T^*$$

or

$$\frac{DT^*}{D(\Pi t^*)} = \nabla^{*2} T^*$$

The dimensionless parameter Π is expressed as follows:

$$\Pi = \frac{k_s}{\rho C_p L U} = \frac{k_s}{\rho C_p L (2\pi N L)} = \frac{k_s}{2\pi \rho C_p N L^2}$$

(7)

where N is rotational velocity in rotations per second. As per the dimensionless equation 6, the heat transfer in the granular bed is conduction controlled. Figure 26 illustrates the variation of the dimensionless average temperatures (T*) of the granular bed (in the impregnator with no baffle and a baffle with 36 cm2 cross-section) with dimensionless time (Πt^*) as a function of either thermal conductivity or heat capacity of the material. The set of five curves (3 for different thermal conductivities and constant heat capacity; and 2 for different heat capacities and constant thermal conductivity) remain very close to each other justifying equation 6 for both the baffled and non-baffled impregnators. Thus, we observe two different groups of curves, one for baffled and one for non-baffled impregnators. The granular beds of baffled impregnator are heated faster in comparison to the non-baffled ones, which is in line to our observation in section 4.5.2. The presence of baffles, which causes a significant difference in the flow and heat transfer of granular material, brings in convective heat transport within the bed.

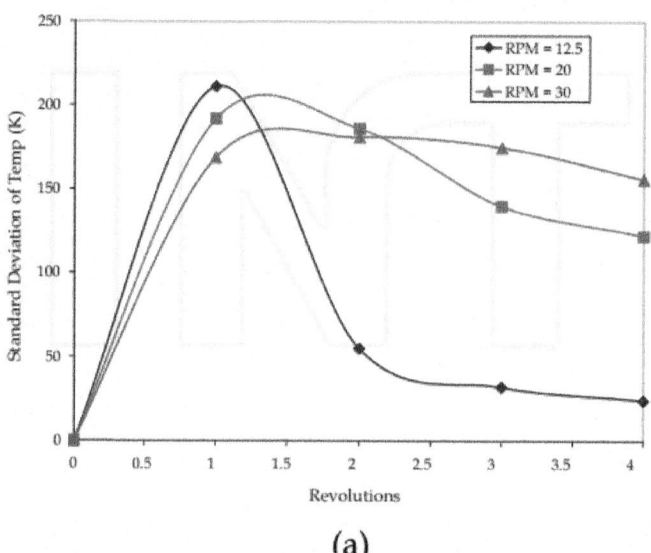

(a)

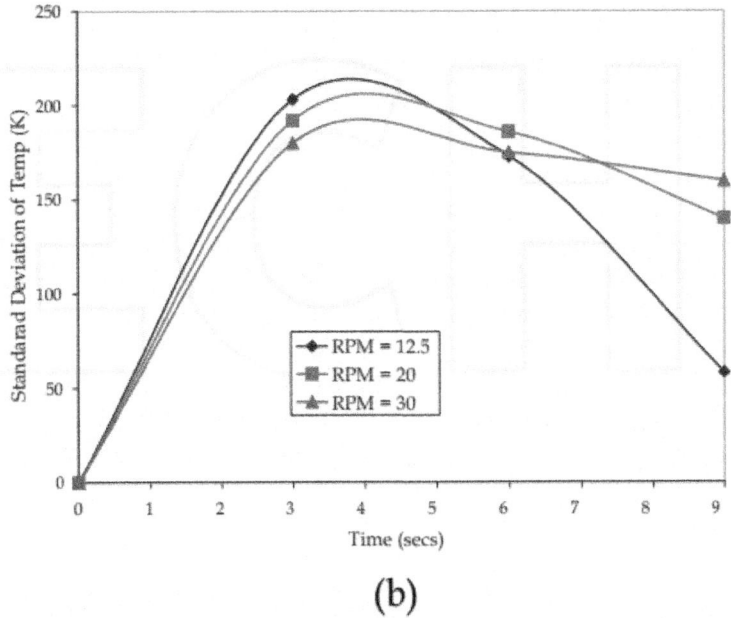

(b)

Figure 25: (a) shows the evolution of standard deviation of the bed temperature versus vessel rotations for different vessel speeds. (b) shows the standard deviation of the bed temperature over real time for different vessel speeds. Lower rotation speed increases the uniformity of bed temperature in both per revolution and real time basis.

CONCLUSIONS

We presented experimentally validated particle dynamics simulations of heat transfer in rotary calciners. Granular flow and heat transport properties of alumina and copper are taken into account in order to develop a fundamental understanding of their effect on calcination performance. Heat transport processes are simulated accounting for initial material temperature, wall temperature, granular heat capacity, granular heat transfer coefficient, and baffle configuration. Simulations and experiments show that the rotation speed has minimal impact on heat transfer. As expected, the material with higher thermal conductivity (alumina) warms up faster in experiments and simulations. We considered various baffle configurations (rectangular and L-shaped flights) in the calciner and their effect on the flow and heat transfer of granular material. Baffles or flights enhance heat transfer and thermal uniformity. L-shaped baffles are more effective than the square shaped baffles. The average wall-particles heat transfer coefficient of the granular system is also estimated from the experimental findings. Particle dynamics simulations were also used to examine heat transfer in granular materials rotated in the impregnators. While particle movement is quite different in the impregnators and calciners, both share similar heat transfer characteristics.

We observe faster heating for materials with higher conductivity and lower heat capacity. Granular cohesion does not affect heat transfer rates. Increasing rotation speed decreases heat transfer and temperature uniformity on a per-revolution basis but the effect disappears on a per-time basis. Impregnator flows also exhibit faster heating of the granular bed for material with higher conductivity and lower heat capacity. Baffles enhance temperature rise in the impregnators. Depending on size, baffles can either increase or decrease heat transfer in the impregnators. Lower rotational speed of the impregnators facilitates heat transfer both on a per-revolution and a per-time basis. A dimensionless parameter Π is derived for the heat transfer in the impregnators, coupling thermal conductivity, heat capacity of the material and the rotational speed of the vessel.

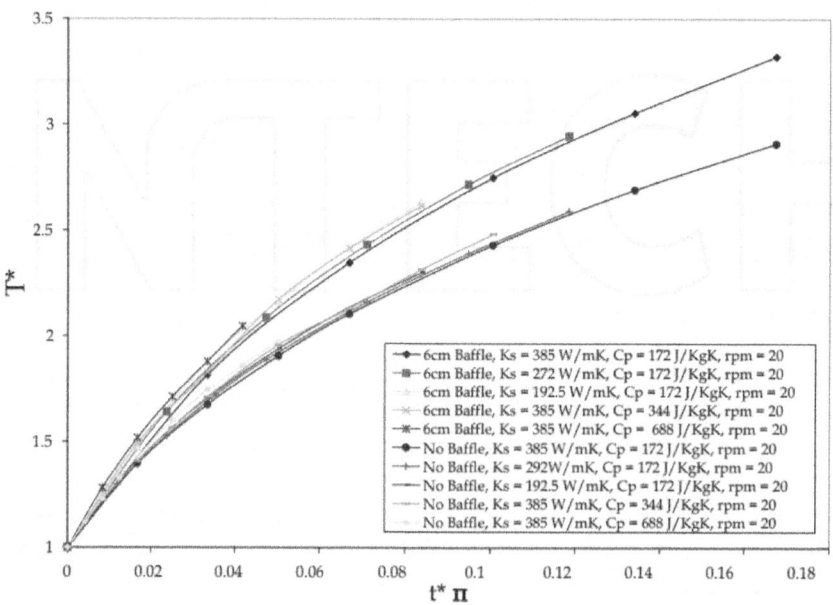

Figure 26: The growth of dimensionless average bed temperature over dimensionless time in the baffled and non-baffled impregnators, for materials with different thermal conductivities (k_s = 192.5, 272, 385 W/mK) and heat capacities (C_p = 172 J/kgK, 344 J/KgK and 688 J/KgK).

Table 1: Parameters employed in DEM simulations

	Notations	Copper	Alumina
Total number of particles	N	8,000-20000	20,000
Radius of the particles	r	2 mm	1.0 mm
Density of the particles	ρ	8900 kg/m3	3900 kg/m3
Specific Heat	C_p	172 J/KgK	875 J/KgK
Thermal Conductivity	k_s	385 W/mK	36 W/mK
Thermal Diffusivity	α	2.5 x 10-7 m2/s	1.1 x 10-5 m2/s
Coefficient of restitution	e		
Particle/particle		0.8	0.8
Particle/wall		0.5	0.5
Normal Stiffness Coefficient	k		
Particle/particle		6000 N/m	6000 N/m
Particle/wall		6000 N/m	6000 N/m
Time step	Δt	1-3 x 10-6 sec	5 x 10-6 sec

ACKNOWLEDGEMENTS

We gratefully acknowledge the support of the Rutgers Consortium on Catalyst Manufacturing Science and Engineering and Pfizer Pharmaceuticals, both grants to FJM and MST. We also acknowledge under graduate students: Myo Kyaw, Dion Zhang and Daniel Carlson for helping with the experiments.

REFERENCES

1. Lee, H. H. (1984). Catalyst preparation by impregnation and activity distribution. Chemical Engineering Science, 39, 859.

2. Lekhal, A., Glasser, B.J., Khinast, J.G. (2001). Impact of drying on the catalyst profile in supported impregnation catalysts. Chemical Engineering Science, 56, 4473.

3. Mickey, H.S., Fairbanks, D.F., (1955). Mechanics of heat transfer to fluidized beds. A.I.CH.E. Journal, 1, 374.

4. Basakov, A.P. (1964) The mechanism of heat transfer between a fluidized bed and surface. International Chemical Engineering, 4, 320.

5. Zeigler, E.N., Agarwal, S. (1969). On the optimum heat transfer coefficient at an exchange surface in a gas fluidized bed, Chemical Engineering Science, 24, 1235.

6. Leong, K.C, Lu., G.Q., Rudolph, V. (2001). Modeling of heat transfer in fluidized bed coating of cylinders. Chemical Engineering Science, 56, 5189.

7. Barletta, M., Simone, G. Tagliaferri, , V. (2005). A FEM model of conventional hot dipping coating process by using fluidized bed. Progress in Organic Coatings, 54, 390.

8. W. Schotte, Thermal conductivity of packed beds. A.I.CH.E Journal, 6, (1960) 63.

9. Sullivan, W.N., Sabersky, R.H. (1975). Heat transfer to flowing granular media. International Journal of Heat and Mass Transfer, 18, 97.

10. Broughton, J., Kubie, J, (1976) A note on heat transfer mechanism as applied to flowing granular media. International Journal of Heat and Mass Transfer, 19, 232.

11. Spelt, J.K, Brennen, C.E., Sabersky, R.H. (1982). Heat transfer to flowing granular material. International Journal of Heat and Mass Transfer, 25, 791.

12. Patton, J.S., Sabersky, R.H., Brennen, C.E. (1987). Convective heat transfer to rapidly flowing granular materials. International Journal of Heat and Mass Transfer, 30, 1663.

13. Buoanno, G., Carotenuto, A. (1996). The effective thermal conductivity of a porous medium with interconnected particles, International Journal of Heat and Mass Transfer, 40, 393.

14. Thomas, B., Mason, M.O., Sprung, R., Liu, Y.A, Squires, A.M. (1998). Heat transfer in shallow vibrated beds. Powder Technology, 99, 293.

15. Cheng, G.J., Yu, A.B., Zulli, Y (1999). Evaluation of effective thermal conductivity from the structure of a packed bed. Chemical Engineering Science, 54, 4199.

16. Wes, G.W.J, Drinkenburg, A.A.H., Stemerding, S. (1976). Heat transfer in a horizontal rotary drum reactor. Powder Technology, 13, 185.

17. Lehmberg, J. , Hehl, M., Schugerl, K. (1977). Transverse mixing and heat transfer in horizontal rotary drum reactors. Powder Technology, 18, 149.

18. Perry, H.R., Chilton, C.H. (1984). C.H. Chemical Engineers' Handbook, McGraw-Hill New York, 6, 11-46.

19. Lybaert, P. (1986). Wall-particle heat transfer in rotating heat exchangers, International Journal of Heat and Mass Transfer, 29, 1263.

20. Boateng, A.A., Barr, P.V. (1998). A thermal model for the rotary kiln including heat transfer within the bed. International Journal of Heat and Mass Transfer, 41, 1929.

21. Le Page, G.P., Tade, M.O, Stone R.J. (1998). Comparitive evaluation of advanced process control techniques for alumina flash calciners. Journal of Process Control, 8, 287.

22. Spurling, R.J., Davidson, J.F., Scott, D.M. (2000). The no-flow problem for granular material in rotating kilns and dish granulators. Chemical Engineering Science, 55, 2303.

23. Sudha, O.S., Chester, A.W., Kowalski, J.A., Beekman, J.W., Muzzio, F.J. (2002). Quantitative characterization of mixing processes in rotary calciners. Powder Technology, 126, 166.

24. Natarajan, V.V.R., Hunt, M.L. (1996). Kinetic theory analysis of heat transfer in granular flows. International Journal of Heat and Mass Transfer, 39, 2131.

25. Michaelidies, E.E. (1986). Heat transfer in particulate flows. International Journal of Heat and Mass Transfer, 29, 265.

26. Ferron, J.R., Singh, D.K. (1991). Rotary kiln transport processes, A.I.CH.E. Journal, 37, 747.

27. Cook, C.A, Cundy, V.A. (1995). Heat transfer between a rotating cylinder and a moist granular bed. International Journal of Heat and Mass Transfer, 38, 419.

28. Cundall, P.A., (1971). A computer model for simulating progressive large-scale movements in blocky rock systems. Proceedings of Symposium International Society of Rock Mechanics, 2, 129.

29. Cundall, P. A., Strack, O. D. L., (1979). A discrete numerical model for granular assemblies. Geotechnique, 29, 47.

30. Dippel, S., Batrouni, G. G., Wolf, D. E., (1996). Collision-induced friction in the motion of a single particle on a bumpy inclined line. Physical Review E, 54, 6845.

31. Luding, S., (1997). Stress distribution in static two dimensional granular model media in the absence of friction. Physical Review E, 55, 4720

32. Thompson, P. S., Grest, G. S., (1991). Granular flow: friction and the dilatancy transition. Physical Review Letters, 67, 1751.

33. Ristow, G. H., Herrmann, H. J., (1994). Density patterns in two-dimensional hoppers. Physical Review E, 50, R5.

34. Wightman, C., Moakher, M., Muzzio, F. J., Walton, O., R., (1998). Simulation of Flow and Mixing of Particles in a Rotating and Rocking Cylinder, A.I.CH.E. Journal, 44, 1226.

35. Shinbrot, T., Alexander, A., Moakher, M., Muzzio, F. J. (1999). Chaotic granular mixing.Chaos, 9, 611.

36. Moakher, M., Shinbrot, T., Muzzio, F. J. (2000). Experimentally validated computations of flow, mixing and segregation of non-cohesive grains in 3D tumbling blenders. Powder Technology, 109, 58.

37. Ristow G. H (1996). Dynamics of granular material in a rotating drum. Europhysics Letters, 34, 263.

38. Walton, O. R., Braun, R. L., (1986). Viscosity, granular-temperature and stress calculations for shearing assemblies of inelastic, frictional disks. Journal of Rheology, 30, 949.

39. Walton, O. R., (1992). Particulate Two-Phase Flow, Butterworth-Heinemann, Boston. Walton, O. R., (1993). Numerical simulation of inclined chute flows of mono disperse, inelastic, frictional spheres. Mechanics of Materials, 16, 239.

40. Sudah, O. S., Arratia, P. E., Alexander, A., Muzzio F. J. (2005). Simulation and experiments of mixing and segregation in a tote blender. A.I.CH.E. Journal, 51, 836.

41. O'Brien, R. W. O., Batchelor, G. K. (1977). Thermal or electrical conduction though granular material, Proc. R. Soc. Lond., 355, 313.

42. Bird, R. B., Stewart, W. E., Lightfoot, E. N. (1960). Transport Phenomena, John Wiley and Sons, New York.

43. Crank, J. (1976). The mathematics of Diffusion, Oxford University Press, UK. Brone, D., Muzzio, F. J. (2000). Enhanced mixing in double-cone blenders. Powder Technology, 110, 179.

44. Muguruma, Y., Tanaka, T., Kawatake, S., Tsuji, Y. (1997). Discrete particle simulation of rotary vessel mixer with baffles. Powder Technology, 93, 261.

Chapter 3

NUMERICAL SIMULATION OF THE UNSTEADY SHOCK INTERACTION OF BLUNT BODY FLOWS

Leonid Bazyma[1], Vasyl Rashkovan[2] and Vladimir Golovanevskiy[3]

[1]National Aerospace University "Kharkov Aviation Institute" Ukraine
[2]National Polytechnic Institute Mexico
[3]Western Australian School of Mines, Curtin University Australia

INTRODUCTION

Supersonic and hypersonic space vehicles are extremely sensitive to aerodynamic resistance. The combination of the two main rocket operation factors, low altitude and high velocity, produces considerable heat flows in the stagnation region of the nose. For this reason, passive heat-transfer analysis under such conditions is very important for understanding and solving rocket operation problems.

The possibility of use of energy supply as the method of the overall control of the airflow is defined in the experimental and theoretical research (Adegren et al., 2001, 2005; Bazyma & Rashkovan, 2005; Tret'yakov et al., 1994, 1996). As an example, reduction of head resistance of a supersonic aircraft can be achieved with the introduction of energy into the contrary incoming flow. On the other hand, supply of energy may be used to minimize negative consequences of the shock-wave interaction when the streamlining of the aerodynamic configurations of the compound form occurs. For example, oblique shock waves, which are distributed from the bow part of an airplane or a rocket, can interact with a bow shock wave of any part of the fuselage construction (tail unit, suspension, hood, diffuser, etc.). In certain cases the shock-wave interaction can result in significant negative and even catastrophic consequences for the aircraft.

The use of a laser to supply energy has been experimentally shown to be a good approach for both general and local flow control. Experimental research (Tret'yakov et al., 1994, 1996) shows that an extensive region of energy supply is realized in the supersonic flow when a powerful optical pulsating discharge is applied, with a thermal wake developing behind the area where the energy is supplied. A cone or hemisphere in the thermal wake, located from 1.0 to 4.0 diameters distance from the focal plane of irradiation from a CO_2 laser, results in a reduction of aerodynamic drag of over a factor of two when a 100-kHz pulse frequency is applied (Tret'yakov et al., 1996). The thermal wake becomes continuous for 10–100-kHz radiation pulses (Tret'yakov et al., 1996).

Theoretical modeling results of the influence of a heat-release pulsating source on the supersonic flow around a hemisphere are presented by Guvernyuk & Samoilov (1997). The explicit total-variation-diminishing (TVD) method in Chakravarthy's) formulae (Chakravarthy & Osher, 1985; Chakravarthy, 1986) is applied in these calculations (Guvernyuk & Samoilov, 1997). In the case of M =3, g =1.4 and a constant deposited energy per unit mass, the aerodynamic load upon the body exhibited a decrease. A pulse repetition rate corresponded to a minimum drag was determined and it was concluded that the use of the pulsating energy supply might be more effective than a constant energy source.

Other results (Georgievskii & Levin, 1988) show that pulse repetition rate, power supplied to the flow and the area into which this energy is supplied all greatly influence both the pressure distribution and the model surface and its flow regimes.

The energy supply parameters such as intensity and heat spot configuration influence the flow re-formation significantly as their combination determines the possibility of either airflow choking in the source (i.e. with the separated wave) or the choke-free flow. This considerably affects the spot properties behind the flow and consequently the stagnation pressure and configuration resistance.

This work presents the results of numerical simulation of the flow around a hemisphere at both the symmetric and asymmetric energy supply into the flow, when the energy supply is realized at 900 angle to the velocity vector of the incoming supersonic airflow.

The two types of the heat spot form considered were: the axis-symmetric spot (i.e. thin disk in the two-dimensional space) and the heat spot of the ellipsoidal form (in the three dimensional space) with its main axis perpendicular to the symmetry axis. The results of the numerical simulation correspond well with the experimental data (Adegren et al., 2001).

Problem definition Let us consider the energy supply from above the sphere and along the airflow at Mach number M_∞ = 3.45 and ratio of specific heats g = 1.4 in the incoming supersonic gas flow, i.e. the same as in (Adegren et al.,

2001). The energy supply scheme is illustrated in Figure 1. Assume that at the time t =0, a pulsating power supply source is initiated in front of the sphere.

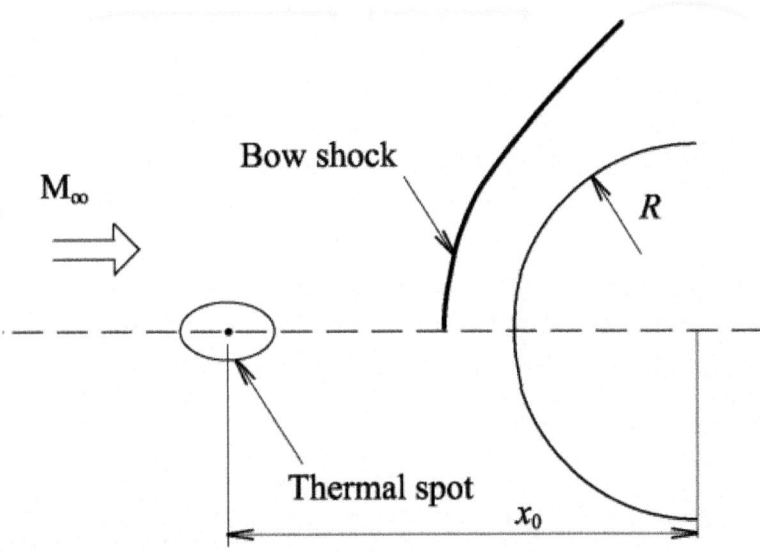

Figure 1: Schematic of energy supply.

The equations of gas dynamics in the cylindrical coordinates, in contrast to (Bazyma & Rashkovan, 2005) are in the form that includes the azimuthal component

$$\frac{\partial \rho r}{\partial t} + \frac{\partial \rho u r}{\partial x} + \frac{\partial \rho \upsilon r}{\partial r} + \frac{\partial \rho \omega}{\partial \varphi} = 0 ;$$

(1)

$$\frac{\partial \rho u r}{\partial t} + \frac{\partial (p + \rho u^2) r}{\partial x} + \frac{\partial \rho u \upsilon r}{\partial r} + \frac{\partial \rho u \omega}{\partial \varphi} = 0 ;$$

(2)

$$\frac{\partial \rho \upsilon r}{\partial t} + \frac{\partial \rho u \upsilon r}{\partial x} + \frac{\partial (p + \rho \upsilon^2) r}{\partial r} + \frac{\partial \rho \upsilon \omega}{\partial \varphi} = p ;$$

(3)

$$\frac{\partial \rho \omega r}{\partial t} + \frac{\partial \rho u \omega r}{\partial x} + \frac{\partial (p + \rho \upsilon^2) r}{\partial r} + \frac{\partial (p + \rho \omega^2)}{\partial \varphi} = 0 ;$$

(4)

$$\frac{\partial per}{\partial t} + \frac{\partial pu(e + p/\rho)r}{\partial x} + \frac{\partial p\upsilon(e + p/\rho)r}{\partial r} + \frac{\partial p\omega(e + p/\rho)}{\partial \varphi} = \rho q r,$$

(5)

where p - pressure; r - density; u, υ, ω - components of the velocity vector on x, r and φ respectively; e – total energy of the mass unit of the gas; q – energy supplied to the mass of the gas by the external source; t – time. The system is completed with the perfect gas equation:

$$p = (\gamma - 1)\rho e.$$

(6)

The energy supply is prescribed the same as in (Bazyma & Rashkovan, 2005; Guvernyuk & Samoilov, 1997):

$$q = W(x,r)\sum_{n=1}^{\infty}\frac{1}{f}\delta\left(t - \frac{n}{f}\right),$$

(7)

where δ – is the Dirak's impulse function; f – pulse repetition rate; W – average mass density of the energy supply. Here, in contrast to (Bazyma & Rashkovan, 2005; Guvernyuk & Samoilov, 1997), W was taken in the form that permits modeling different shapes of heat spot at the asymmetric energy supply:

$$W = W_0\left(\frac{p_\infty}{\rho_\infty}\right)^{3/2}\frac{1}{R}\exp\left(-\frac{k_1(r\cos\varphi)^2 + k_2(x - x_0)^2 + k_3(r\sin\varphi)^2}{L^2}\right),$$

(8)

where W_0, k_1, k_2, k_3 and L are constants defining deposited energy density and thermal spot shape.

NUMERICAL METHOD

Similar to (Bazyma & Rashkovan, 2005), the solution of the system of equations (1-4) was conducted using the Godunov´s method (Godunov, 1976). The 110´60 and the 110´60×32 grids were used for the axis-symmetric and the three-dimensional problems respectively. In both cases the grid was designed with deepening of the nodes near the body or in the areas of energy supply (i.e. the incoming flow disturbance areas).

The calculations were performed using the same finite difference scheme of the first-order approximation as that used by (Godunov, 1976). Computational grid used in calculations is shown in Figure 2.

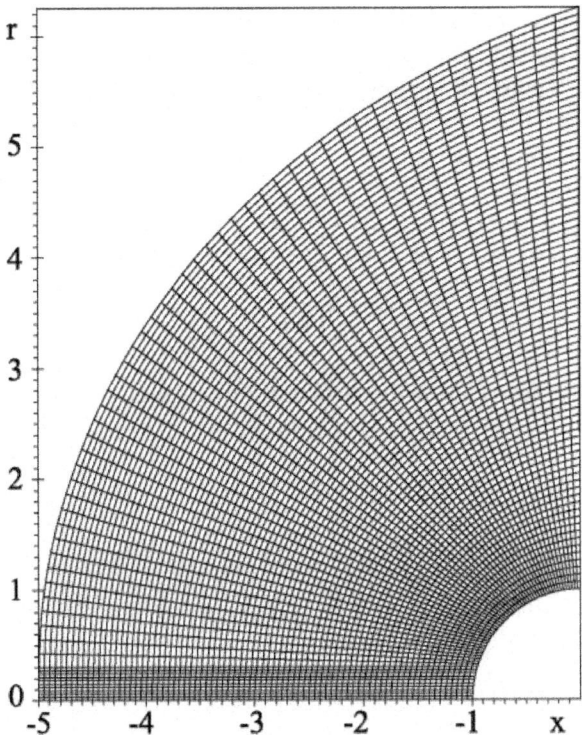

Figure 2: Computational grid.

The necessary and sufficient condition for stability is that the permissible spacing in time τ must satisfy the inequality

$$\frac{\tau}{\tau_x} + \frac{\tau}{\tau_y} \leq 1$$

(9)

resulting from a stability study of Godunov's difference scheme realized on the system of non-stationary acoustic equations on a uniform rectangular (or parallelogram) grid. Here, τ_x and τ_y are the time spacing of the one-dimensional scheme. Physically τ_x and τ_y are mean time intervals, in which waves appearing at the break decomposition on the cell boundary reach the neighboring boundaries:

$$\tau_x = \frac{\Delta x}{\max(u+a, a-u)}, \tau_y = \frac{\Delta y}{\max(v+a, a-v)},$$

(10)

where a is the velocity of the sound.

The stability condition so given is extended to the quasi-linear equations of gas dynamics. Calculations show that this condition (9) provides the necessary stability. Nevertheless, this condition is usually used with right-hand side less than one. In our work, the time step was chosen from cell to cell according to the stability condition as follows:

$$\tau_{n-\frac{1}{2},m-\frac{1}{2}} = \left(\frac{\tau_x \tau_r}{\tau_x + \tau_r}\right)_{n-\frac{1}{2},m-\frac{1}{2}}, \bar{\tau} = \min_{n,m} \tau_{n-\frac{1}{2},m-\frac{1}{2}},$$

(11)

In "k" space its value with respect to time is calculated using the spacing "k+1" as:

$$\tau^{k+1} = K\bar{\tau}^k,$$

(12)

where K is a safety factor similar in meaning to the Courant number.

In order to use dimensionless values, we make use of the following equalities:

$$r = \bar{r}R, \quad x = \bar{x}R, \quad t = \bar{t}R / a_\infty, \quad f = \bar{f}a_\infty / R, \quad a = \bar{a}a_\infty,$$

$$u = \bar{u}a_\infty, \quad \upsilon = \bar{\upsilon}a_\infty, \quad \omega = \bar{\omega}a_\infty \quad \rho = \bar{\rho}\rho_\infty, \quad p = \bar{p}\rho_\infty a_\infty^2, \quad W = \bar{W}a_\infty^3 / R,$$

(13)

where a∞ the velocity of the sound of the incident flow. In the following text, we have omitted bars above the dimensionless values r, x, t, f, a, u, u, r, p, W.

The boundary conditions are defined similar to those in (Guvernyuk & Samoilov, 1997). On the surface of the body and along the symmetry axis, solid-wall inviscid boundary conditions were applied. Along the external inflow boundary, undisturbed freestream conditions were utilized. On the downstream outflow boundary, extrapolation of the flow quantities from the adjacent internal boundary was performed.

The initial data in calculations without energy deposition corresponded to the dimensionless parameters of the incident stream:

$$p = p_\infty = 1/\gamma, \quad \rho = \rho_\infty = 1, \quad u = u_\infty = M_\infty, \quad \upsilon = 0, \quad \omega = 0,$$

(14)

where g is the ratio of specific heats.

Subsequent solutions of the flow field about a hemisphere with energy deposition were initialized using the flow field about a hemisphere solution without energy deposition. The solutions were advanced in time until the average flow conditions were stabilized. Preliminary supersonic flow calculations around the hemisphere were conducted in (Bazyma & Rashkovan,

2005) to confirm the adequacy of our numerical scheme (see Fig. 3). Data reported in (Guvernyuk & Samoilov, 1997), where the explicit TVD method of Chakravarthy's formulation (Chakravarthy & Osher, 1985; Chakravarthy, 1986) was used, showed good correspondence in all the observed flow regimes.

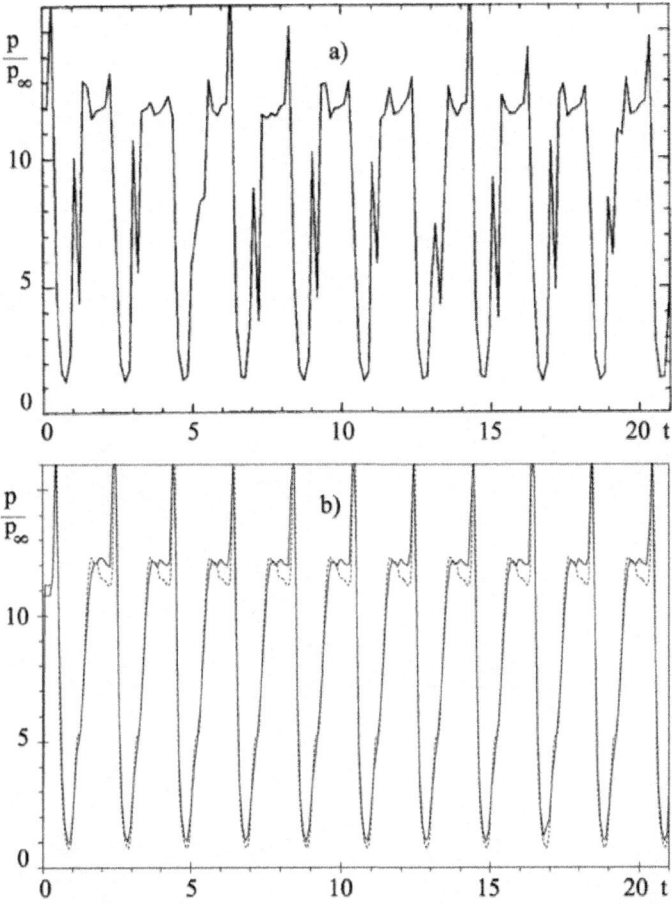

Figure 3: Pressure at the hemisphere stagnation point versus dimensionless time at the pulse repetition rate f=0.5 (spherical heat spot; W0 =20, x0 = -3.5, L = 0.5): a) results of (Guvernyuk & Samoilov, 1997); b) results of (Bazyma & Rashkovan, 2005) (——— 110×60 grid; - - - - - 219×119 grid).

Shown in Figure 3 is the dependence of the pressure at the stagnation point while flowing around the hemisphere (for $M_\infty = 3$, g = 1.4) on the dimensionless time at pulse repetition rate f = 0.5. Results are shown both from (Guvernyuk & Samoilov, 1997) (a) and from the our previous work (b). As can be seen from Figure 3, the main pulsation parameters (i.e. period and amplitude) and their character obtained in our work correspond well with the data of (Guvernyuk & Samoilov, 1997). However, while the resolution of the numerical scheme used in (Guvernyuk & Samoilov, 1997) is somewhat higher the scheme used in our previous work (Bazyma & Rashkovan, 2005) allows simulation of the quasi-stationary pulsation process.

Shown in Figure 4a is the flow visualization near the hemisphere under the influence of the pulsating thermal source (spherical heat spot; $W_0 = 20$, $x_0 = -3.5$, $L = 0.5$). These results were obtained in (Guvernyuk & Samoilov, 1997), and compared with the analogous data of the (Bazyma & Rashkovan, 2005) (Figure 4b). As can be seen in Figure 4, bow shock wave standoff distance, formed recirculation zones and the flow in general reported in (Guvernyuk & Samoilov, 1997) and derived in our work are in good agreement.

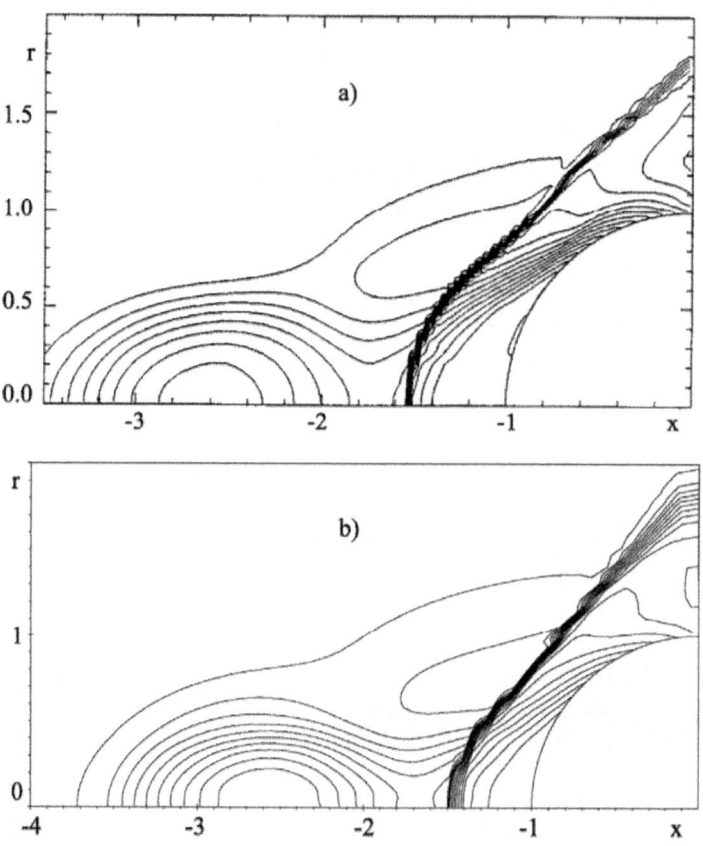

Figure 4: Mach number isolines while flowing around the hemisphere with supersonic gas flow at pulse repetition rate f=2 (t=11.2): a) results of (Guvernyuk & Samoilov, 1997); b) results of (Bazyma & Rashkovan, 2005), 110×60 grid.

The grid resolution study was also conducted, with test calculations for hemisphere and cavity hemisphere carried out using the 219×119 grid. The 219×119 grid was obtained through twice the 110×60 grid spacing reduction. Minimum surface cell spacing values were 0.024 for the 110×60 grid (reduced to the sphere radius) and 0.012 for the 219×119 grid.

A comparison of the results derived with the use of the 110×60 (continuous line) and 219×119 grids can be seen in Figure 3b above. As the resolution of the 219×119 (dotted line) grid is higher than that of the 110×60 grid, the solution derived with the use of the 219×119 grid marked out some peculiarities of the pressure change on the compression stage. These peculiarities correspond to the solution in (Guvernyuk & Samoilov, 1997) as well; however they were not seen through the grid applied. It is worth noting that solution difference obtained with the 110×60 and 219×119 grids is rather small for the hemisphere with energy deposition.

The details of the numerical scheme, along with test examples, are given in (Godunov, 1976) and (Bazyma & Kholyavko, 1996; Bazyma & Rashkovan, 2006).

RESULTS AND DISCUSSION

When calculating the three-dimensional problem, the energy supply modeled was at 900 angle to the velocity vector of the incoming flow. The two types of the heat spot form considered were: the axis-symmetric spot (i.e. thin disk) and the heat spot of the ellipsoidal form with the main axis perpendicular to the symmetry axis. The dimensionless parameters of the undisturbed contrary flow are assumed as the initial data in calculations without power supply. The Table 1 shows the list of operational parameters for the facility used in the wind tunnel experiments (Adegren et al., 2001). This tunnel is a basic blowdown tunnel with an exhaust into atmospheric pressure.

Table 1: Operating Parameters for the Rutgers Mach 3.45 Supersonic Wind Tunnel.

Mach Number	3.45
Operating Stagnation Pressure	1.4 MPa
Typical Stagnation Temperature	290 K
Mass Flow Rate	9.8 Kg/s
Total Run Time	1.8 minutes
Test Area Cross Section	15 cm x 15 cm
Test Area Length	30 cm

General calculation procedure

The energy supplied to the mass unit of gas is prescribed in the form

$$q = \gamma^{-3/2} W_0 \exp\left(-\frac{k_1(r\cos\varphi)^2 + k_2(x-x_0)^2 + k_3(r\sin\varphi)^2}{L^2}\right) t'\delta(t-t').$$

(15)

Here $x_0 = -3.0$ (the energy is supplied at the distance of one diameter of the sphere from its surface, (Adegren et al., 2001)), $t* = f--1$ at the pulse frequency $f = 0.00068$ (that corresponds to the frequency 10 H_z, (Adegren et al., 2001)). The form of the heat spot is defined by the parameters L, k_1, k_2, k_3. The value $L = 0.01$ is fixed in all the calculations; the values k_1, k_2, k_3 are being variated, that permitted to obtain the heat spot dimensions characteristic for the experiment (Adegren et al., 2001) (the volume of the heat spot is evaluated approximately from 1 to 3 mm3; The sphere radius in the experiment is 12.75 mm). Thus, for example, at $k_1 = k_2 = k_3$ one can obtain a spherical heat spot.

The parameter W_0 is being varied in the range $0.19 - 1.75$ that in total with the selection of values k_1, k_2, k_3 provides the change of the energy density in the impulse that is provided in the experiment (Adegren et al., 2001) (13 mJ/pulse/1±0.5mm³, 127 mJ/pulse/1.3±0.7mm³, and 258 mJ/pulse/3±1mm³).

The bow shock stand-off distance for the undisturbed model at Mach 3.45 was calculated and compared to the Lobb (Lobb, 1964) approximation to Van Dyke's (Van Dyke, 2003) shock stand-off model. The model predicted the stand-off distances within 3 percent of the calculated distances. The model for shock stand-off distance is given as

$$\Delta = 0.41 D \frac{(\gamma-1)}{(\gamma+1)} \frac{M_\infty^2 + 2}{M_\infty^2},$$

(16)

where, Δ is the stand-off distance, D is the sphere diameter, and $M_\infty = 3.45$ is the freestream Mach number.

Symmetric energy supply

Naturally, obtaining the heat spot form similar to that used in the experiment (Adegren et al., 2001) for the two-dimensional case is impossible. However, with the heat spot size small compared to the size of the streamlined body (i.e. sphere) and low values of the energy density supplied to the incoming flow obtaining some similarity can be expected. For the spheroidal heat spot form (i.e. its axis of rotation coincides with the axis of rotation of the streamlined body), the character of the pressure change in the critical point of the body is sufficiently close (at the stage of compression and the first phase of expansion) to that obtained in the experiment (Adegren et al., 2001) for the value of energy supplied in the pulse in the order of 13mJ/1mm$_3$ (see Figure 5).

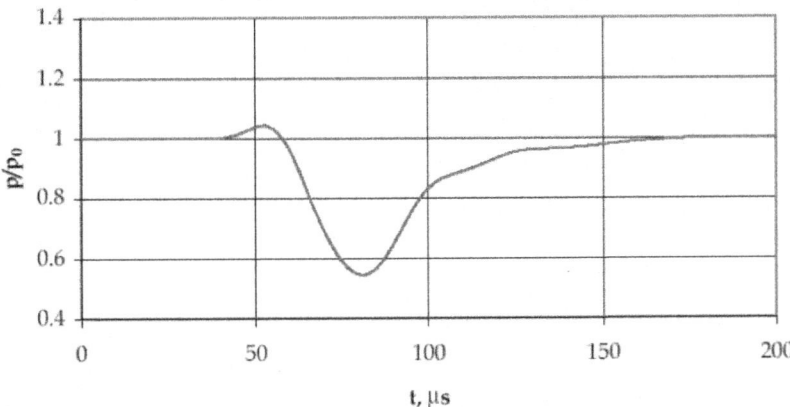

Figure 5: Pressure variation at the critical point of the sphere versus time: $W_0 = 0.19$, $f = 0.00068$, ellipsoid heat point; $k_1 = k_3 = 0.016$, $k_2 = 0.39$.

Increasing the spot size and the energy supply density will not allow obtaining the conditions fully adequate to those of the physical experiment. At the same time, varying these two parameters (i.e. the size and configuration of the spot on the one hand and the energy supply density on the other hand) allow the character of the pressure change in the critical point of the body adequate to the experiment by the higher values of the energy supply density (127 mJ/pulse/1.3±0.7mm³, and 258 mJ/pulse/3±1mm³) to be obtained.

Character of pressure variation in the critical point of the sphere versus time for other values of k_1, k_2, and k_3 providing smaller volume of the heat spot (i.e. by the factor of 2 approximately) but twice the energy supply density (i.e. $W_0 = 0.38$) is shown in Figure 6, curve 1. It is worth noting that the character of the pressure variation in both cases is similar. Curves 2 and 3 in Figure 6, with a considerable difference in pressure amplitude at the compression stage, are obtained for the form of the heat spot that is the flattened axissymmetric disk with its radius comparable with the radius of the sphere.

As in the experiment (Adegren et al., 2001; refer figure 20) the time history of recorded pressure at centerline location of sphere surface for the three energy levels shows a common behavior comprised of an initial pressure rise, expansion, compression and transient decay.

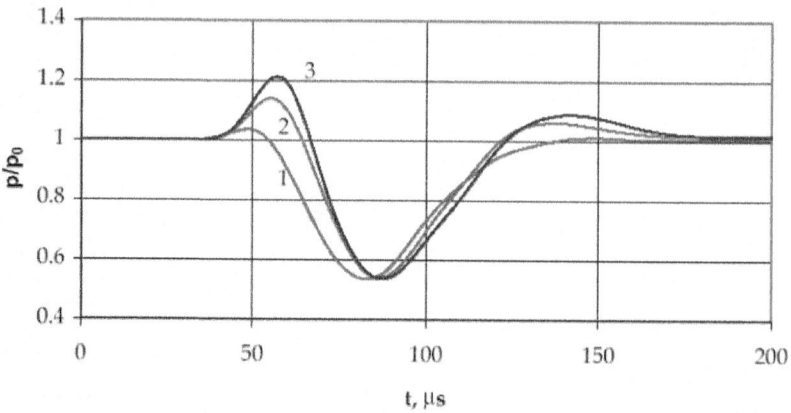

Figure 6: Pressure variation at the critical point of the sphere versus time: 1 – W0 = 0.38, f = 0.00068, ellipsoid heat point; k1 = k3 = 0.016, k2 =0.56; 2 – W0 = 0.38, f = 0.00068, disk heat point; k1 = k3 = 0.00045, k2 =0.089;3 – W0 = 0.38, f = 0.00068, disk heat point; k1 = k3 = 0.00023, k2 =0.082.

The expansion, compression and transient decay are similar to the ideal gas Euler simulations of (Georgievskii & Levin, 1993) for the interaction of a thermal spot with a sphere at Mach 3. The expansion lowers the surface pressure at the centerline by 40%. However, the initial compression phase observed in our research and in the experiment (Adegren et al., 2001) was not noted by (Georgievskii & Levin, 1993).

The interaction of the thermal spot, with the bow shock (Figure 7-9, t = 40-90 microseconds) causes a blooming of the bow shock (due to the lens effect of the thermal spot). This behavior is consistent with the simulations of Georgievski and Levin (Georgievskii & Levin, 1993).

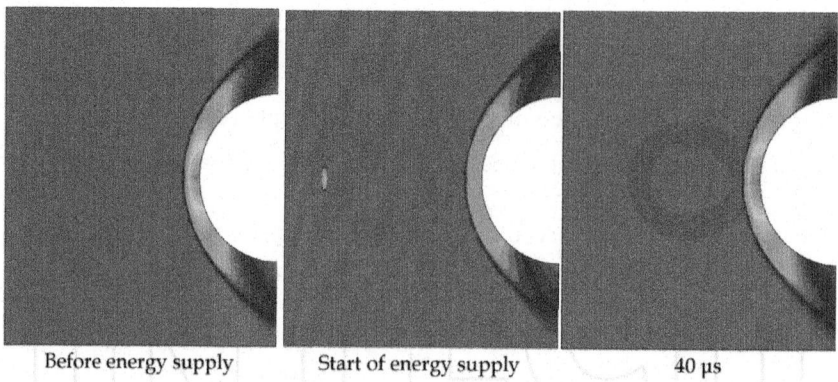

Before energy supply   Start of energy supply   40 µs

Figure 7: Time histories of pressure isolines: $W_0 = 0.38$, $f = 0.00068$, disk heat point; $k_1 = k_3 = 0.00045$, $k_2 = 0.089$.

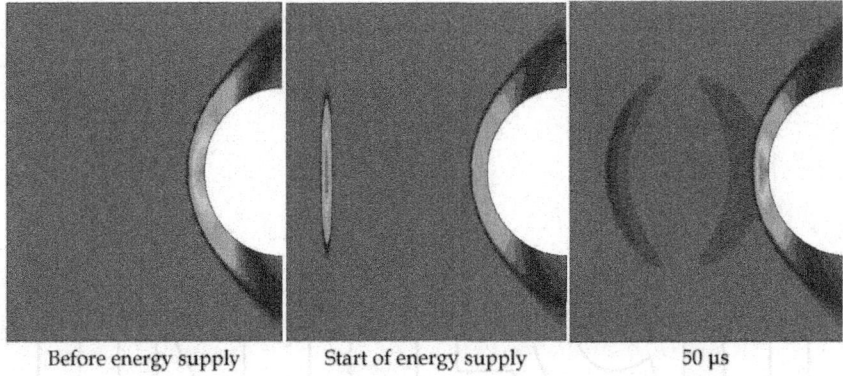

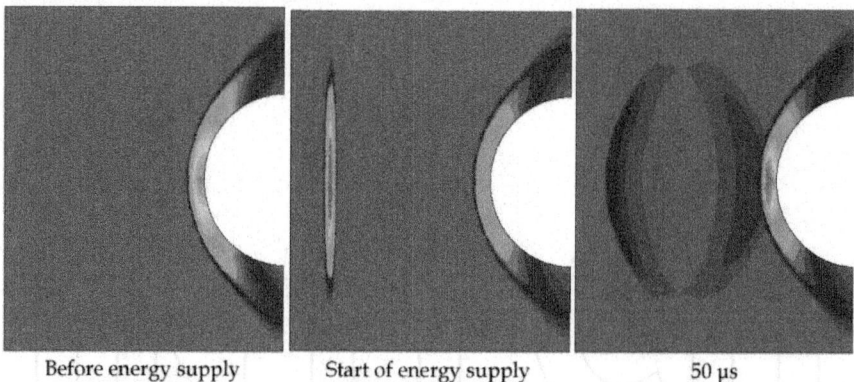

Figure 8: Time histories of pressure isolines: $W_0 = 0.38$, $f = 0.00068$, disk heat point; $k_1 = k_3 = 0.00045$, $k_2 = 0.089$.

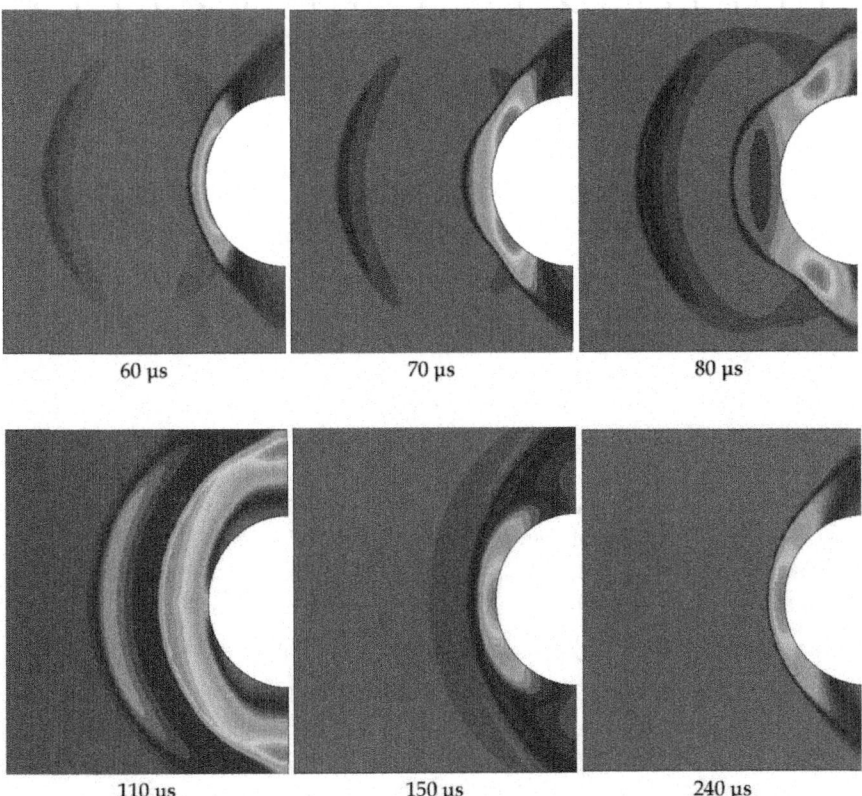

60 µs   70 µs   80 µs

110 µs   150 µs   240 µs

Figure 9: Time histories of pressure are isolines: $W_0 = 0.38$, $f = 0.00068$, disk heat point; $k_1 = k_3 = 0.00023$, $k_2 = 0.082$.

Naturally, with these conditions satisfied the energy supply density was even less than in the experiment (Adegren et al., 2001). However, the character of the pressure variation at the stage of compression and expansion was similar to that obtained in the experiment for energy supply densities of 127 mJ/pulse/$1.3 \pm 0.7 mm^3$ and 258 mJ/pulse/$3 \pm 1 mm^3$.

Asymmetric energy supply

Pressure variation at the critical point of the sphere, relevant to the corresponding pressure obtained before the heat influence, versus time after the energy supply is shown in Figure 10.

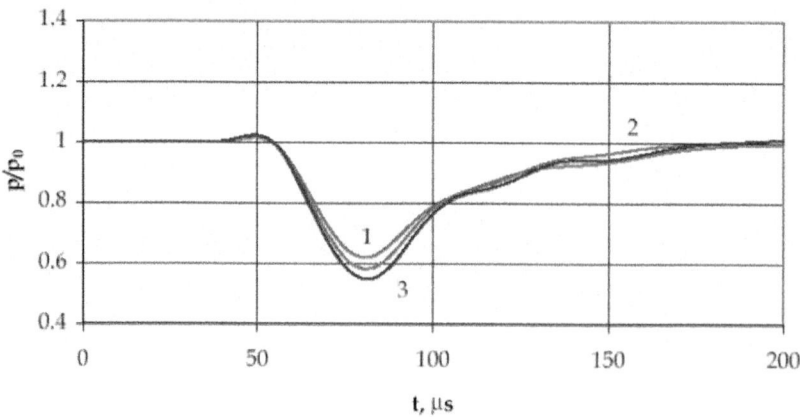

Figure 10: Pressure variation at the critical point of the sphere versus time: $1 - W_0$ = 0.5, f = 0.00068, ellipsoid heat point, $k_2 = k_3 = 0.67$, $k_1 = 0.007$; $2 - W_0 = 1.5$, f = 0.00068, ellipsoid heat point, $k_2 = k_3 = 1$, $k_1 = 0.06$; $3 - W_0 = 1.75$, f = 0.00068, ellipsoid heat point, $k_2 = k_3 = 1$, $k_1 = 0.17$.

Figure 11 shows the pressure history for the asymmetric energy deposition. The Mach number in the contrary flow is M=3.45. The distance from the heat spot to the sphere was equal to one diameter of the sphere. The zone of the energy supply was an ellipsoid with the volume in the order of 1mm3 with its rotational axis perpendicular to the incoming flow velocity vector. The process of interaction of the heat track and the sphere contained two stages: short stage of compression and long stage of expansion. The obtained results are in good agreement with the experimental data (Adegren et al., 2001), both quantitatively and qualitatively.

Figure 12 shows the fields of the equal pressure on the sphere surface at 500 μs time after the heat influence (here the pressure is related to the contrary flow pressure). As can be seen, by this time the symmetrical vortex areas on the surface of the sphere still exist. It is worthwhile to point out that the flow rotation velocity is sufficiently high; with its maximum value reaching 0.3 of the contrary flow sound velocity (see Figure 13).

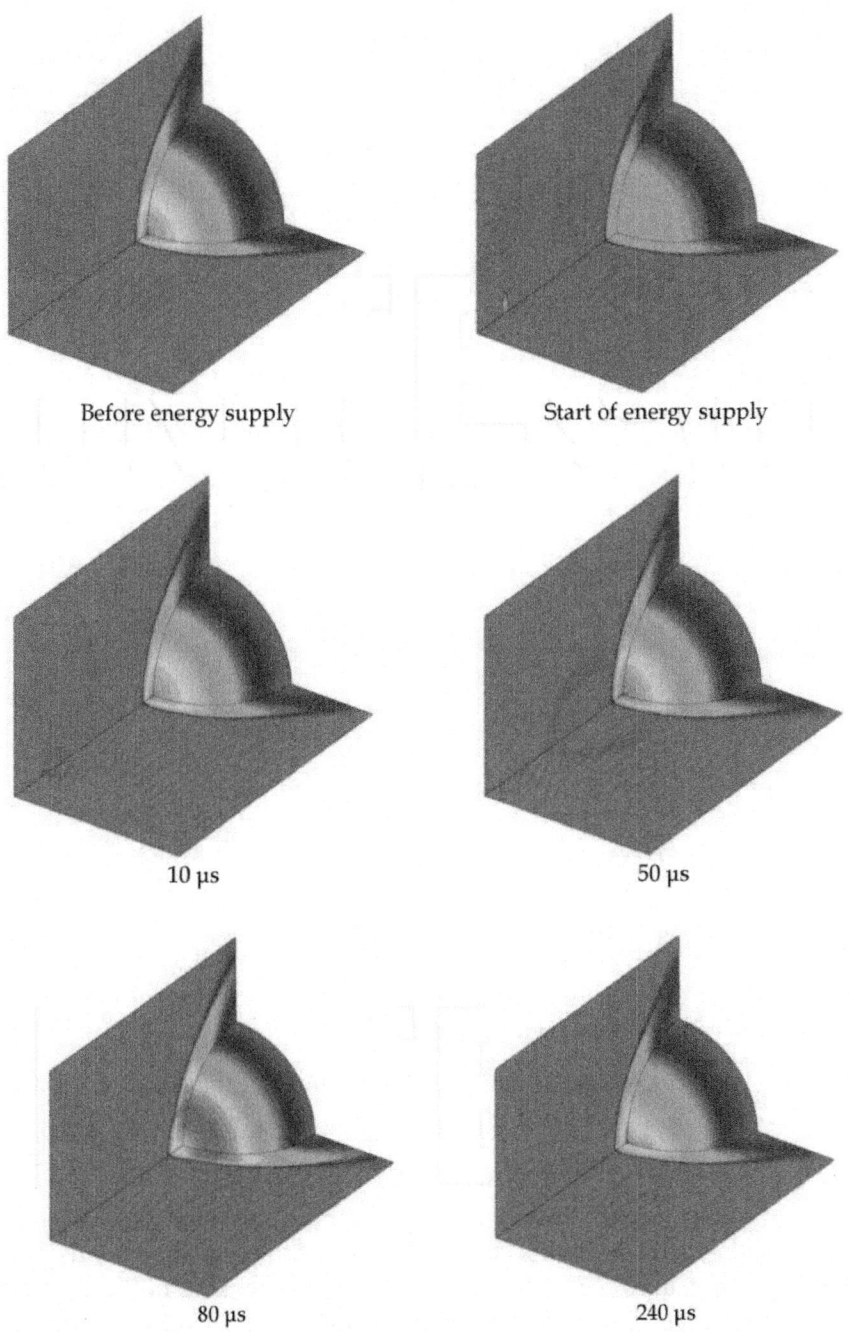

Before energy supply

Start of energy supply

10 μs

50 μs

80 μs

240 μs

Figure 11: Time histories of pressure isolines: $W_0 = 0.5$, $f = 0.00068$, ellipsoid heat point, $k_2 = k_3 = 0.67$, $k_1 = 0.007$.

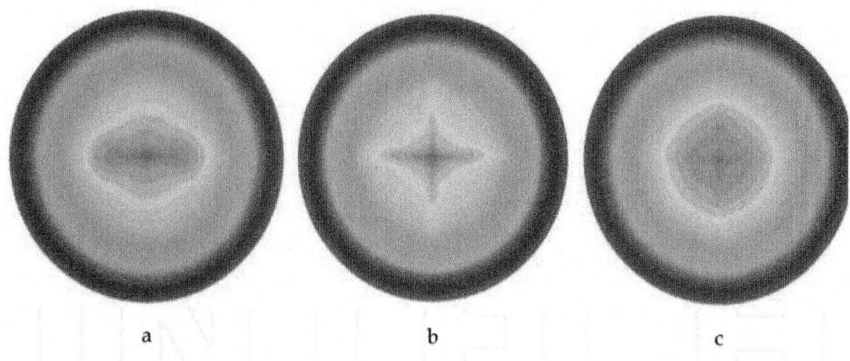

Figure 12: Pressure on the sphere surface at 500 μs time after the energy supply:

a - $W_0 = 0.5$, $f = 0.00068$, ellipsoid heat point, $k_2 = k_3 = 0.67$, $k_1 = 0.007$;

b – $W_0 = 1.5$, $f = 0.00068$, ellipsoid heat point, $k_2 = k_3 = 1$, $k_1 = 0.06$;

c - $W_0 = 1.75$, $f = 0.00068$, ellipsoid heat point, $k_2 = k_3 = 1$, $k_1 = 0.17$.

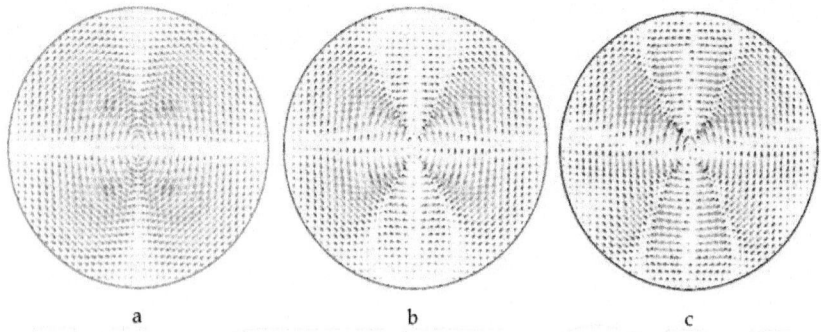

Figure 13: The azimuthal velocity component (in the vector form) on the sphere surface at 500μs time after the energy supply:

a - $W_0 = 0.5$, $f = 0.00068$, ellipsoid heat point, $k_2 = k_3 = 0.67$, $k_1 = 0.007$;

b – $W_0 = 1.5$, $f = 0.00068$, ellipsoid heat point, $k_2 = k_3 = 1$, $k_1 = 0.06$;

c - $W_0 = 1.75$, $f = 0.00068$, ellipsoid heat point, $k_2 = k_3 = 1$, $k_1 = 0.17$.

The energy supply parameters, i.e. the heat spot configuration and the energy supply intensity, influence the flow reconstruction significantly as their combination defines the possibility of either flow choking in the source (i.e. with the separated wave) or the chokefree flow. This considerably affects the track properties behind the flow and consequently the braking pressure and the configuration resistance. Energy supply into the incoming airflow allows the loads caused by the shock-wave influence on the construction elements of aircraft to be minimized.

CONCLUSION

The results of our research can be summarized as follows:

- An initial flow compression stage during the lens and blooming interaction process has been observed for both symmetric and asymmetric energy supply conditions;

- A 40% decrease in surface pressure during the 40 microsecond thermal spot interaction time was observed for the case of a sphere in Mach 3.45 flow with varying levels of energy deposition upstream of the bow shock for both symmetric and asymmetric energy supply conditions;

- For the case of symmetric energy supply, the process of heat track and sphere interaction has been found to consist of three stages i.e. an initial compression, expansion, compression and transient decay regardless of the energy input density magnitude;

- For asymmetric energy supply, the process of heat track and sphere interaction has been found to consist of two stages i.e. a short stage of compression followed by a long stage of expansion;

- Thermal spot shape and intensity of energy supply influence reorganization of flow, with their combination affecting properties of the track and stagnation pressure in both symmetric and asymmetric energy supply cases.

The application of energy deposition for local flow control requires low power in terms of the energy deposition into the flow. This low power requirement could potentially translate into small, low weight energy generation systems utilizing optical lasers or electric arc units etc. for effective and efficient flow control. This warrants consideration of the use of energy supply as effective means for solution of local issues of supersonic flow of various aircraft.

ACKNOWLEDGMENT

The authors would like to express their gratitude to Professor Alexander V. Gaydachuk of the National Aerospace University "Kharkov Aviation Institute" for his input and guidance during the manuscript preparation.

REFERENCES

1.   Adegren, R.; Elliot, G.; Knight D.; Zheltovodov, A. & Beutner, T. (2001). Energy deposition in supersonic flows, AIAA Paper N2001–0885, 2001

2. Adelgren, R.; Yan, H.; Elliott, G.; Knight D.; Beutner, T. & Zheltovodov, A. (2005). Control on Edney IV Interaction by Pulsed Laser Energy Deposition, AIAA J., Vol. 43, No. 2, pp. 256-269, ISSN 0001-1452

3. Bazyma, L. & Rashkovan, V. (2005). Stabilization of Blunt Nose Cavity Flows Using Energy Deposition. Journal of Spacecraft and Rockets, Vol.42, No.5, (September-October 2005), pp. 790-794, ISSN 0022-4650

4. Bazyma, L. & Kholyavko V.I. (1996). A modification of Godunov's finite difference scheme on a mobile grid. Computational Mathematics and Mathematical Physics, Vol. 36, No.4, pp. 525-532, ISSN 0965-5425

5. Bazyma, L. & Rashkovan, V. (2006). Separation Flow Control by the Gas Injection Contrary Supersonic Stream, AIAA J., Vol. 44, No. 12, pp. 2887-2895, ISSN 0001-1452

6. Chakravarthy, S. & Osher, S. (1985). New Class of High Accuracy TVD Schemes for Hyperbolic Conservation Laws, AIAA Paper 85-0363

7. Chakravarthy, S. (1986). The Versality and Reliability of Euler Solvers Based on HighAccuracy TVD Formulations," AIAA Paper 86-0243

8. Georgievskii, P. & Levin, B. (1988). Supersonic Flow Around Bodies in the Presence of External Heat Pulsed Sources, Technical Physics Letters, Vol. 14, No. 8, pp. 684–687 (ISSN 1063-7850)

9. Georgievski, P. & Levin, V. (1993). Unsteady Interaction of a Sphere with Atmospheric Temperature In homogeneity at Supersonic Speed, Akademiya Nauk SSSR, Izvestiya, Mekhanika Zhidkosti i Gaza, No. 4, June, pp. 174-183, ISSN 0568-5281

10. Godunov, S. Ed(s). (1976). Numerical Solution of Multidimensional Problems in Gas Dynamics, Nauka, Moscow (in Russian)

11. Guvernyuk, S. & Samoilov, A. (1997). Control of supersonic flow around bodies by means of a pulsed heat source. Technical Physics Letters, Vol.23, No.5, pp. 333-336, ISSN 1063-7850

12. Lobb, R.K. (1964). Experimental Measurement of Shock Detachment Distance on Spheres Fired in Air at Hypervelocities, In The High Temperature Aspects of Hypersonic Flow, Nelson, WC (ed.), pp. 519-527, Pergamon Press, New York, NY

13. Tret'yakov, P.; Grachev, G.; Ivanchenko, A.; Krainev, V.; Ponomarenko, A. & Tishenko, V. (1994). Optical breakdown stabilization in the supersonic argon flow. PhysicsDoclady, Vol.39, No.6, pp. 415-416, ISSN 1063-7753

14. Tret'yakov, P.; Garanin, A.; Grachev, G.; Krainev, V.; Ponomarenko, A.; Tishenko, V. & Yakovlev, V. (1996). Control of supersonic flow around

bodies by means of highpower recurrent optical breakdown. Physics-Doclady, Vol.41, No.11, pp. 566-567, ISSN 1063-7753

15. Van Dyke, M.D. (2003). The Supersonic Blunt-Body Problem - Review and Extension. AIAA J., Vol. 41, No. 7, pp. 265-276, ISSN 0001-1452

Chapter 4

INVERSE ANALYSIS APPLIED TO MUSHY STEEL RHEOLOGICAL PROPERTIES TESTING USING HYBRID NUMERICAL-ANALYTICAL MODEL

Miroslaw Glowacki
AGH University of Science and Technology Poland

INTRODUCTION

Integrated casting and rolling technologies are most recent and very efficient way of hot strip production. More and more companies all over the world are able to manage such processes. The mentioned technologies ensure huge reduction of rolling costs, very high product quality and low investment costs. Computer simulation is of vital importance to the development of "know how" theory for these processes. The lack of publications concerning mechanical properties and behavior of steels simultaneously subjected to both plastic deformation and solidification was the inspiration for the investigation. This also necessitated the development of an appropriate mathematical model of mushy-steel deformation. The contribution summarizes the results of the author's recent theoretical research concerning the computer simulation of mushy steel published in recent years in well-known journals and book chapters [Glowacki, 2006; Glowacki at al., 2010; Glowacki & Hojny, 2006, 2009; Hojny & Glowacki, 2008, 2009a, 2009b, 2011; Hojny at al., 2009].

As an example of a company providing the integrated casting and rolling technologies one can mention the plant located in Cremona, Italy which develops the new methods of steel strip manufacturing. They are called Inline Strip Production (ISP) and Arvedi Steel Technology (AST) processes and are characterized by very high temperature allowed at the mill entry. The instant rolling of slabs which leave the casting machine allows for the utilization of the heat stored in the strips during inline casting. Both the mentioned technologies ensure huge reduction of rolling forces and their details are usually classified. The development of "know how" theory for the semi-solid steel rolling technology requires numerical modeling.

The development of appropriate mathematical models is limited by the lack of thermal and mechanical properties concerning mushy steels deformation in temperature range which is close to solidus line. The work presented in the current contribution is an attempt to cover the gap providing a proposition of a hybrid numerical analytical model of semi-solid steel deformation. The mathematical modeling of steel deformation in semi-solid state, as well as experimental work in this field, is innovative topics regarding the very high temperature range deformation processes. Tracing the related papers published in the past 10 years one can find many dealings with experimental results for non-ferrous metals tests (Kang & Yoon, 1997; Koc at al., 1996; Kopp at al., 2003; Sang Yong at al., 2001; Zhao at al., 2006). The first results regarding steel deformation at extra high temperature were presented during last few years (Li, 2005; Seol, 1999, 2002). Most of the problems concerning semi-solid steel testing are caused by the very high level of steel liquidus and solidus temperatures in comparison with non-ferrous metals. The deformation tests for non-ferrous metals are much easier. The rising abilities of thermo-mechanical simulators enable investigation of steel samples and as a result both computer simulation and the development of new, very high temperature rolling technologies like Arvedi ISP and AST processes. The lack of mathematical models describing the steel behavior in the last phase of solidification with simultaneous plastic deformation was the inspiration of the investigation described in the proposed book chapter.

The main goal of the chapter is to present problems of theoretical work leading to the development of a methodology of very high temperature testing of steel samples while their central parts are still mushy. In such conditions the deformation of samples is strongly inhomogeneous and all the well-known methods of yield stress curve examination fail due to significant barrelling of the sample. Although the investigation concerned both physical tests and dedicated simulation system, the author sacrifices the contribution to the hybrid model which is the heart of the system. With the help of inverse analysis it allows for the right interpretation of deformation tests providing data regarding the mushy steel rheological properties.

PHYSICAL BASIS AND CHARACTERISTIC FEATURES OF STEEL DEFORMED AT VERY HIGH TEMPERATURE

The rolling equipment for the ISP process allows for reduction of initial mould strip thickness from 74 mm to 55 mm during liquid core reduction process. The region of maximum strip temperature for a high reduction mill is located in the strip centre and varies from 1220 °C to 1375 °C depending on the casting speed. The main benefits of the technology are: inverse temperature gradient, good product quality, very low level of heating energy consumption, up to 20 times lower water consumption in comparison to traditional rolling, low level of installed mill power, compact rolling equipment layout, no need for tunnel furnace and very low investment costs. The AST technology is a result of further development of ISP into a real endless process and the benefits of its application are even greater. The whole reduction process is running in one rolling mill consisting of 5 or 7 stands, which can reduce the strip thickness from 55÷70 mm to 0.8 mm. The maximum temperature of the strip occurs in central region of its cross-section and varies from 1340 °C to 1420 °C depending on the casting speed. This suggests that the central region of the strand subjected to the rolling is still mushy.

The main benefits of the new very high temperature technologies are significantly lower rolling forces and very favorable temperature field inside the steel plate. However, certain problems arise which are specific for this kind of metal treatment. The central parts of slabs are mushy and the solidification is not yet finished while the deformation is in progress. This results in changes in material density and occurrence of characteristic temperatures having great influence on the plastic behavior of the material (Senk, 2000; Suzuki, 1988). The nil strength temperature (NST), strength recovery temperature (SRT), nil ductility temperature (NDT) and ductility recovery temperature (DRT) have effect on steel plastic behavior and limit plastic deformation. The Nil Strength Temperature (NST) is the temperature level at which material strength drops to zero while the steel is being heated above the solidus temperature. Another temperature associated with NST is the Strength Recovery Temperature (SRT). At this temperature the cooled material regains strength greater than 0.5 N/mm2. Nil Ductility Temperature (NDT) represents the temperature at which the heated steel loses its ductility. The Ductility Recovery Temperature (DRT) is the temperature at which the ductility of the material (characterised by reduction of area) reaches 5% while it is being cooled. Over this temperature the plastic deformation is not allowed at any stress tensor configuration.

Significant changes of density and lack of data regarding material's thermal and mechanical properties are vital problems of the modeling. They have great influence on steel rheology and heat transfer. An issue of great importance is the lack of strain-stress relationships, which in the temperature range above 1400 °C strongly depend on the density and are very temperature sensitive. It is not easy to run isothermal tests that could be the source of the computation of yield stress function parameters for such high temperatures. There are also some problems with the interpretation of tests results.

Density is very important for plastic behavior of mushy steel plates. It varies with temperature and depends on the cooling rate. The solidification process causes non-uniform density distribution in the controlled volume resulting in non-uniform deformation and heat conduction. There are three main factors causing density changes: solid phase formation, thermal shrinkage and movement of liquid particles inside the solid skeleton. The density plays an important role in both mechanical and thermal solutions. The contribution sheds some light on the physical problems but it focuses on the axial symmetrical computer model, which ensures the right simulation of mushy steel samples deformation reflecting the physical requirements. The presented model fills the gap in modeling of plastic behavior of semi-solid steels.

HYBRID NUMERICAL-ANALYTICAL MODEL OF MUSHY STEEL DEFORMATION

Testing of steels at temperature higher than 1400 °C is difficult due to deformation instability and risk of sample damage during experiment. Such experiments do not assure the strain homogeneity and cannot be interpreted using traditional methods. Appropriate interpretation of the results is possible only with the help of a computer aided engineering system. The contribution reports a new model underlying such a system developed by the author's team. Together with GLEEBLE physical simulator equipped with high temperature module the code allows for investigation of properties of semi-solid steel.

The numerical solver is the less visible yet very powerful kernel of the system. It is based on a thermal-mechanical model with variable density. The mechanical part of the model is a hybrid variational solution with analytical mass conservation condition constraining the velocity field components. The accuracy of the proposed solution is very good due to negligible volume loss guaranteed by the analytical form of the mass conservation condition. This is important for materials with variable density and is not captured by classical solutions. Analytical condition eliminates problems with unintentional specimen volume changes caused by application of numerical methods. The existing, physical changes of steel density in the mushy zone have influence on real variations of controlled volume.

On the other hand numerical errors can be a source of volume loss which interferes with real changes. This effect is very undesirable in modelling of thermal-mechanical behaviour of steel in temperature range characteristic for the (transformation of state of aggregation). The mentioned mechanical and thermal parts of the mathematical model of the process are supported by a third one, i.e. the density changes model. The mechanical part is responsible for the strain, strain rate and stress distribution in a controlled volume.

THERMAL PART OF THE MODEL

Heat exchange between solid metal and environment, and its flow inside the metal is controlled by a number of factors. During phase change two additional phenomena have to be taken into account. Note that in the process of deformation of steel at temperature of liquid to solid phase transition there are two sources of heat changes. On the one hand heat is generated due to the state transformation. On the other hand it is secreted as a result of plastic deformation. In addition, steel density variations also cause changes of body temperature.

Thermal solution has a major impact on simulation results, since the temperature has strong effect on remaining variables. This is especially evident if the specimen temperature is close to solidus line when the body consist of both solid and semi-solid regions. In such case the affected phenomena are: plastic flow of solid and mushy materials, stress evolution and density changes. The theoretical temperature field is a solution of Fourier-Kirchhoff equation with appropriate boundary conditions.

The most general form of the Fourier-Kirchhoff equation in any coordinate system can be written in operator form as follows:

$$\nabla^T (\Lambda \nabla T) + Q = c_p \rho \left(v^T \nabla T + \frac{\partial T}{\partial \tau} \right) \tag{1}$$

where T is the temperature distribution in the controlled volume and Λ denotes the symmetrical second order tensor called heat transformation tensor. In case of thermal inhomogeneity the whole tensor has to be considered. Q represents the rate of heat generation (or consumption) due to the phase transformation, due to plastic work done and due to electric current flow (resistance heating of the sample is usually applied). Finally C_p describes the specific heat, ρ the steel density, v the velocity vector of specimen particles and τ the elapsed time. The heat transformation tensor consists of a set of anisotropic heat transformation coefficients and can be given in a form:

$$\Lambda = \begin{pmatrix} \lambda_{xx} & \lambda_{xy} & \lambda_{xz} \\ \lambda_{yx} & \lambda_{yy} & \lambda_{yz} \\ \lambda_{zx} & \lambda_{zy} & \lambda_{zz} \end{pmatrix} \tag{2}$$

In the case of anisotropic bodies, the solution is carried out locally, and the axes of coordinate system are oriented in accordance with the principal directions of the thermal conductivity. In this case all off-diagonal components of the heat transformation tensor are zeros $(\lambda_{ij} = 0, i \neq j)$ and equation (2) becomes:

$$\Lambda = \begin{pmatrix} \lambda_{xx} & 0 & 0 \\ 0 & \lambda_{yy} & 0 \\ 0 & 0 & \lambda_{zz} \end{pmatrix}$$

(3)

Furthermore for a thermally isotropic material $\lambda_{xx} = \lambda_{yy} = \lambda_{zz} = \lambda$ and tensor of the heat transformation can be written in the index notation can be as:

$$\Lambda_{ij} = \lambda \delta_{ij}$$

(4)

where δ_{ij} is the Kronecker delta.

The temperature of samples compressed in axially-symmetric process can be determined by solving the appropriate form of Fourier-Kirchhoff equation. Here the equation will be expressed in the cylindrical coordinate system, which is a natural choice for the cylindrically-shaped samples. It takes following differential form:

$$\frac{1}{r}\frac{\partial}{\partial r}\left(r\lambda_r \frac{\partial T}{\partial r}\right) + \frac{1}{r}\frac{\partial}{\partial \theta}\left(\frac{1}{r}\lambda_\theta \frac{\partial T}{\partial \theta}\right) + \frac{\partial}{\partial z}\left(\lambda_z \frac{\partial T}{\partial z}\right) + Q = \rho c_p \frac{\partial T}{\partial \tau}$$

(5)

The assumption of axial symmetry can be considered appropriate for the tensile and compression tests of steel in semi-solid state in all physically stable cases. It is invalid only for failed experiments. The symmetry simplifies the model by implying identical temperature distribution at any axial sample cross-section. This results in the equation:

$$\frac{\partial T}{\partial \theta} = 0$$

(6)

Equation (5) can be further simplified if the heat properties of the medium are assumed isotropic. By calculating the differentials in equation (5) and using equation (6) we get the following form of Fourier-Kirchhoff equations for isotropic, axially-symmetric heat flow:

$$\lambda\left(\frac{\partial^2 T}{\partial r^2} + \frac{1}{r}\frac{\partial T}{\partial r} + \frac{\partial^2 T}{\partial z^2}\right) + Q = \rho c_p \frac{\partial T}{\partial \tau}$$

(7)

Equation (7) needs to be solved with appropriate initial and boundary conditions. The initial conditions relate to cases of non-stationary heat exchange. Most solutions use Cauchy condition which assume the known a priori temperature distribution at time τ_0: $T_\Omega(\tau_0) = f_\Omega$. In a particular (but often adopted) case the temperature is assumed to be constant throughout the considered area $T_\Omega(\tau_0) = T_0 = const$.

Boundary conditions have more complex nature and relate to all cases of heat transfer and describe the spatial aspect of the heat exchange. The considered continuous medium changes its temperature though convection, radiation, conduction, or a combination of these phenomena. Theoretical solutions of the problem are generally subject to one or more boundary conditions. Combined Hankel's boundary conditions have been adopted for the presented model. The conditions for axially-symmetrical problem can be written in form of a differential equation:

$$\lambda r \frac{\partial T}{\partial n} + \alpha(T - T_0) + q = 0$$

(8)

In equation (8) T_0 is the distribution of border temperature, q describes the heat flux through the boundary of the deformation zone, a is the heat transfer coefficient and n is a vector which is normal to the boundary surface. More details concerning the problem can be found in (Glowacki, 1996). Equation (7) subject to condition (8) defines the problem of temperature evolution during the whole process of heating and deformation of the samples.

Note that (7) is a spatiotemporal equation. The solution of such equations is difficult because in general case the temperature is a function of both location (r,z) and time τ.

$$T = T(r, z, \tau)$$

(9)

In addition, the used boundary conditions, appropriate for the cooling or heating of the sample are also described by differential equation. For that reason equation (7) is solved in a two-step process (Zienkiewicz at al., 2005):

- the corresponding steady-state equation is solved. After FEM discretization this yields a matrix algebraic equation,
- the solution obtained in the first step is then adapted to non-steady-state conditions using a transient discretization of the time variable.

Thermal model for steady-state heat flow process

The Fourier-Kirchhof equation (7) for the steady heat flow can be written as:

$$\lambda\left(\frac{\partial^2 T}{\partial r^2} + \frac{1}{r}\frac{\partial T}{\partial r} + \frac{\partial^2 T}{\partial z^2}\right) + Q = 0$$

(10)

Application of finite element method for solving problems of heat flow requires a functional. Equation (10) together with the boundary conditions given by equation (8) needs to be expressed in a variation setting.

Consider the problem of optimizing the general form of the heat flux power functional.

$$\chi = \int_V f(r, z, T, T_r, T_z) \, dV + \int_S \left(qT + \frac{1}{2}\alpha(T - T_0)^2\right) dS$$

(11)

where f is a function of position, temperature and temperature gradient:

$$T_r = \frac{\partial T}{\partial r}; \qquad T_z = \frac{\partial T}{\partial z}$$

(12)

This function is specified in the relevant domain V with the boundary S. Let us consider a small variation of the functional (11):

$$\delta\chi = \int_V \left(\frac{\partial f}{\partial T}\delta T + \frac{\partial f}{\partial T_r}\delta T_r + \frac{\partial f}{\partial T_z}\delta T_z\right) dV + \int_S [q\delta T + \alpha(T - T_0)\,\delta T]dS$$

(13)

that can be rewritten as:

$$\delta\chi = \int_V \delta T \left[\frac{\partial f}{\partial T} - \frac{\partial}{\partial r}\left(\frac{\partial f}{\partial T_r}\right) - \frac{\partial}{\partial z}\left(\frac{\partial f}{\partial T_z}\right)\right] dV + \int_S \delta T\left(q + \alpha(T - T_0) + l_r\frac{\partial f}{\partial T_r} + l_z\frac{\partial f}{\partial T_z}\right) dS$$

(14)

where l_r and l_z are the direction cosines of normal to the outer surface with respect to or and oz-axes, respectively.

A necessary condition for the functional (11) to reach extreme value for a given function is for the variation $\delta\chi$ to be equal to 0. Since equation (14) must be satisfied for any variation δT, the expressions in brackets have to be zero at an extreme:

$$\frac{\partial}{\partial r}\left(\frac{\partial f}{\partial T_r}\right) + \frac{\partial}{\partial z}\left(\frac{\partial f}{\partial T_z}\right) - \frac{\partial f}{\partial T} = 0$$

(15)

for the entire volume V and

$$l_x \frac{\partial f}{\partial T_x} + l_z \frac{\partial f}{\partial T_z} + q + \alpha(T - T_0) = 0$$

(16)

for its boundary S. Can therefore be concluded that if one satisfy the equations (15) and (16) than the functional (11) reaches an optimum. Both of these formulations are equivalent. The above reasoning is the solution of so called Euler problem. In the presented particular case the appropriate form of the function f is as follows:

$$f = r\left[\frac{1}{2}\lambda(T_r^2 + T_z^2) - QT\right]$$

(17)

where T_r and T_z are given by relationships (12). In this case the equations (15) and (16) can be written as follows:

$$\lambda\left(\frac{\partial^2 T}{\partial r^2} + \frac{1}{r}T_r + \frac{\partial^2 T}{\partial z^2}\right) + Q = 0$$

$$\lambda r\frac{\partial T}{\partial n} + q + \alpha(T - T_0) = 0$$

(18)

The presented reasoning shows that the assumption of steady-state heat flow leads to equations (18). The first of them is identical with the equation (10), and the second to boundary condition (8). Thus, according to the Euler reasoning, the solution of equation (10) satisfying the boundary condition (8) is the functional extremal:

$$\chi = \int_V r\left\{\frac{1}{2}\lambda\left[\left(\frac{\partial T}{\partial r}\right)^2 + \left(\frac{\partial T}{\partial z}\right)^2\right] - QT\right\}dV + \int_S \left(qT + \frac{1}{2}\alpha(T - T_0)^2\right)dS$$

(19)

Optimization of the functional (19) in the domain of discrete functions is based on replacement of the continuous real function of the temperature distribution T (r, z) by their discrete counterparts. In the proposed solution the finite element method was used for that purpose. The discretization of the control volume was done accordingly. The temperature distribution function was discretized according to the formula:

$$T(r,z) = \mathbf{n}^T(r,z)\,\mathbf{T}$$

(20)

where n (r, z) is a vector of the shape function and T is a nodal temperature vector. After substituting (20) and its derivatives to (19) it takes the discrete form:

$$\chi = \int_V r \left\{ \frac{1}{2} \lambda \left[\left(\frac{\partial \mathbf{n}^T}{\partial r} \mathbf{T} \right)^2 + \left(\frac{\partial \mathbf{n}^T}{\partial z} \mathbf{T} \right)^2 \right] - Q \mathbf{n}^T \mathbf{T} \right\} dV + \int_S \left(q \mathbf{n}^T \mathbf{T} + \frac{1}{2} \alpha (\mathbf{n}^T \mathbf{T} - T_0)^2 \right) dS$$

(21)

From the mathematical point of view, equation (21) no longer defines a functional, but a function of many variables. Nevertheless hereinafter it still will be referred to as a functional. Its derivative with respect to T is given as follows:

$$\frac{\partial \chi}{\partial \mathbf{T}} = \int_V r \left[\lambda \mathbf{T}^T \left(\frac{\partial \mathbf{n}}{\partial r} \frac{\partial \mathbf{n}^T}{\partial r} + \frac{\partial \mathbf{n}}{\partial z} \frac{\partial \mathbf{n}^T}{\partial z} \right) - Q \mathbf{n}^T \right] dV + \int_S (q \mathbf{n}^T + \alpha (\mathbf{T}^T \mathbf{n} - T_0) \mathbf{n}^T) dS$$

(22)

Equation (22) one can written in matrix form as:

HT + p = 0   (23)

where matrix **H** and vector **p** have shapes:

$$\mathbf{H} = \int_V r \lambda \left(\frac{\partial \mathbf{n}}{\partial r} \frac{\partial \mathbf{n}^T}{\partial r} + \frac{\partial \mathbf{n}}{\partial z} \frac{\partial \mathbf{n}^T}{\partial z} \right) dV + \int_S \alpha \mathbf{n} \mathbf{n}^T dS$$

$$\mathbf{p} = - \int_V r Q \mathbf{n} \; dV - \int_S (\alpha T_0 - q) \, \mathbf{n} \; dS$$

(24)

The system of linear equations (23) can be solved using standard methods of linear algebra. This yields the discrete temperature vector **T**.

Thermal model for non-steady-state heat flow

For the non-steady-state heat flow equation (7) has to be used instead of equation (10). A derivation similar to the one for the steady-state flow and the same space discretization lead to formulation of discrete form of functional equivalent to equation (7). It is analogous to functional (21).

$$\chi = \int_V r \left\{ \frac{1}{2} \lambda \left[\left(\frac{\partial \mathbf{n}^T}{\partial r} \mathbf{T} \right)^2 + \left(\frac{\partial \mathbf{n}^T}{\partial z} \mathbf{T} \right)^2 \right] - \left[Q - \rho c_p \frac{\partial}{\partial \tau} (\mathbf{n}^T \mathbf{T}) \right] \mathbf{n}^T \mathbf{T} \right\} dV +$$

$$+ \int_S \left(q \mathbf{n}^T \mathbf{T} + \frac{1}{2} \alpha (\mathbf{n}^T \mathbf{T} - T_0)^2 \right) dS$$

(25)

Differentiation of functional (25) with respect to **T** leads to relation which is similar to (22).

$$\frac{\partial \chi}{\partial \mathbf{T}} = \int_V r\left[\lambda \mathbf{T}^T\left(\frac{\partial \mathbf{n}}{\partial r}\frac{\partial \mathbf{n}^T}{\partial r} + \frac{\partial \mathbf{n}}{\partial z}\frac{\partial \mathbf{n}^T}{\partial z}\right) - \left(Q - \rho c_p \frac{\partial \mathbf{T}^T}{\partial \tau}\mathbf{n}\right)\mathbf{n}^T\right]dV +$$
$$+ \int_S (q\mathbf{n}^T + \alpha(\mathbf{T}^T\mathbf{n} - T_0)\mathbf{n}^T)dS \tag{26}$$

The system (26) can be written in a matrix form analogous to equation (23):

$$\mathbf{HT} + \mathbf{C}\frac{\partial \mathbf{T}}{\partial \tau} + \mathbf{p} = \mathbf{0} \tag{27}$$

where $\mathbf{H}$ and $\mathbf{p}$ are matrices given by (24), and $\mathbf{C}$ can be expressed as:

$$\mathbf{C} = \int_V \rho c_p \mathbf{n} \ \mathbf{n}^T dV \tag{28}$$

An assumption of linear temperature change in very short time interval $\Delta\tau$ and application of weighted Galerkin's residual method leads to an equation which is a discrete (with respect to time) counterpart of equation (27).

$$\overline{\mathbf{H}}\mathbf{T}_{i+1} + \overline{\mathbf{p}} = \mathbf{0} \tag{29}$$

Matrix $\overline{\mathbf{H}}$ and vector $\overline{\mathbf{p}}$ in equation (29) are described by the following relations:

$$\overline{\mathbf{H}} = \left(2\mathbf{H} + \frac{3}{\Delta\tau}\mathbf{C}\right)$$
$$\overline{\mathbf{p}} = \left(\mathbf{H} - \frac{3}{\Delta\tau}\mathbf{C}\right)T_i + 3\mathbf{p} \tag{30}$$

Equation (29) can be used to compute the vector of nodal temperatures $\mathbf{T}_{i+1}$ after a time step $\Delta\tau$ (i.e. at $\tau = \tau_{i+1} = \tau_i + \Delta\tau$) provided that initial value $\mathbf{T}_i$ for $\tau = \tau_i$ is known.

MECHANICAL MODEL

A mathematical model of the compression process is based on the theory of plastic flow (Chakrabarty, 2006). The principle of the upper assessment (Bower, 2010), calculus of variations (Adhikari, 1998), approximation theory and optimization methods (Findaeisen at al., 1980 ; Nocedal & Wright 2006) and numerical methods for solving partial differential equations (Evans 1988; Polyanin, & Zaitsev, 2004; Pinchover & Rubinstein, 2005), including the finite element method (Zienkiewicz at al., 2005) were used. The following assumptions were established:

- deformation and stress state are axial-symmetrical,
- deformed material is isotropic but inhomogeneous,
- the material behaviour is rigid-plastic - the relationship between the stress tensor and strain rate tensor is calculated according to the Levy-Mises flow law, which is given as:

$$\sigma_{ij} - \frac{1}{3}\sigma_{kk}\delta_{ij} = \frac{2}{3}\frac{\sigma_p}{\dot{\varepsilon}_i}\dot{\varepsilon}_{ij}$$

(31)

Rigid-plastic model was selected due to its very good accuracy at the strain field during the hot deformation and sufficient correctness of calculated deviatoric part of the stress field. Moreover, the elastic part of each stress tensor component is very low at temperatures close to solidus line and can in practice be neglected in calculations of strain distribution. The limits for plastic metal behavior are defined according to Huber-Mises-Hencky yield criterion:

$$\sigma_{ij}\sigma_{ij} = 2\left(\frac{\sigma_p}{\sqrt{3}}\right)^2$$

(32)

In equations (31) and (32) σ_{ij} denotes the stress tensor components, σ_{kk} represents the mean stress, δ_{ij} is the Kronecker delta, σ_p indicates the yield stress, $\dot{\varepsilon}_i$ is the effective strain rate, and $\dot{\varepsilon}_{ij}$ denotes strain rate tensor components. The components are given by an equation:

$$\dot{\varepsilon}_{ij} = \frac{1}{2}\left(\nabla_i v_j + \nabla_j v_i\right)$$

(33)

In cylindrical coordinate system $Or\theta z$ the solution is a vector velocity field defined by the distribution of three coordinates $v = (v_r, v_\theta, v_z)$. The field is a result of optimization of a power functional, which can be written in general form as the sum of power necessary to run the main physical phenomena related to plastic deformation. Due to the axial-symmetry of the sample the velocity field the circumferential component of the velocity field can be neglected and the functional is usually formulated as:

$$J[v] = W = \dot{W}_\sigma + \dot{W}_\lambda + \dot{W}_f$$

(34)

Component $\dot{W}_\sigma$ occurring in equation (34) represents the plastic deformation power, $\dot{W}_\lambda$ is the power which is a penalty for the departure from mass conservation condition, $\dot{W}_f$ denotes the friction power and $v = (v_r, v_z)$ describes the reduced velocity field distribution. Rigid-plastic formulation of metal deformation problem requires the condition of mass conservation in the deformation zone. In case of solids and liquids with a constant density, this condition can be simplified to the incompressibility condition. Such a condition is generally satisfied with sufficient accuracy during the optimization of functional (34). In most solutions a slight, but noticeable loss of volume is observed. The loss is caused by incomplete fulfillment of the incompressibility condition imposed on the solution in numerical form. It is negligible in case of traditional computer simulation of deformation processes although in some embodiments more accurate methods are used to restore the volume of metal subjected to the deformation. Unlike this case the density of semi-solid materials varies during the deformation process and these changes result in a physically reasonable change in the volume of a body having constant mass. The size of the volume loss due to numerical errors is comparable with changes caused by fluctuation in the density of the material.

A further problem specific to the variable density continuum is power $\dot{W}_\lambda$, which occurs in functional (34). It is used in most solutions and has a significant share of total power. Even when the iterative process approaches the end, this power component is still significant, especially if the convergence of the optimization procedures is insufficient. In case of discretization of the deformation area (e.g. using the finite element method) if one focuses solely on the $\dot{W}_\lambda$ a number of possible optimal solutions appear. They are related to a number of possible directions of movement of discretization nodes providing the volume preservation of the deformation zone. Each of these solutions creates a local optimum for $\dot{W}_\lambda$ power and thus for the entire functional (34). This makes it difficult to optimize because of lack of uniform direction of fall of total power which leads to global optimum. The material density fluctuation causes further optimization difficulties, resulting from additional replacement of incompressibility condition with a full condition of mass conservation

The proposed solution requires high accuracy in ensuring the incompressibility condition for the solid material or mass conservation condition for the semi-solid areas. This approach stems from the fact that the errors resulting from the breach of these conditions can be treated as a volume change caused by the steel density variation in the semi-solid zone. High accuracy solution is required also due to large differences in yield stress for the individual subareas of the deformation zone. In the discussed temperature range they appear due to even slight fluctuations in temperature. In presented solution the second component of functional (34) is left out and mass conservation condition is given in analytical form constraining the radial (V$_r$)and longitudinal (V$_z$)velocity field components. The functional takes the following shape:

$$J[v] = \dot{W}_\sigma + \dot{W}_t$$

(35)

In case of functional (35) the numerical optimization procedure converges faster than the one for functional (34) due to the reduced number of velocity field parameters (only radial components are optimization parameters) and the lack of numerical form of mass conservation condition. The accuracy of the proposed hybrid solution is higher also due to negligible volume loss caused by numerical errors which is very important for materials with variable density.

As mentioned before the solution of the problem is a velocity field in cylindrical coordinate system in axial-symmetrical state of deformation. Optimization of metal flow velocity field in the deformation zone of semi-variational problem requires the formulation according to equation (35). The radial velocity distribution $v_r(r, \theta, z)$ and the longitudinal one $v_z(r, \theta, z)$ are so complex that such wording in the global coordinate system poses considerable difficulties. These difficulties are the result of the mutual dependence of these velocities. Therefore the basic formulation will be written for the local cylindrical coordinate system $Or\theta z$ with a view to the future discretization of deformation area using one of the dedicated methods. In addition one will find that the deformation of cylindrical samples is characterized by axial symmetry. As demonstrated by experimental studies conducted using semi-solid samples the symmetry may be disturbed only as a result of unexpected leakage of liquid phase.

Such experiments, however, are regarded as unsuccessful and not subject to numerical analysis. Establishment of the axial symmetry, which except in cases of physical instability can be considered valid also for the process of compression or tensile test of semi-solid samples, allows one to simplify the model because of the identical strain distribution at any axial sample cross-section. Considerations will therefore be carried out in Orz coordinates for the sample cross-sectional using one of the planes containing the sample axis. Components of power functional given by (35) have been formulated in accordance with the general theory of plasticity by relevant equations. The plastic power for the deformation zone having volume of V is given by the subsequent relation:

$$\dot{W}_\sigma = \int_V \sigma_i \dot{\varepsilon}_i \, dV$$

(36)

where σ_i is the effective stress and $\dot{\varepsilon}_i$ denotes the effective strain. The plastic deformation starts when the rising effective stress reaches yield stress limit σ_p $(\sigma_i = \sigma_p)$ according to yield criterion given by equation (32). Effective strain occurring in equation (36) is calculated on the basis of the strain tensor components $\dot{\varepsilon}_{ij}$ according to following relationship:

$$\dot{\varepsilon}_i = \sqrt{\frac{2}{3} \dot{\varepsilon}_{ij} \dot{\varepsilon}_{ij}}$$

(37)

The components are given by equation (33). For axial-symmetrical case the strain has a form:

$$\begin{pmatrix} \dfrac{\partial v_r}{\partial r} & 0 & \dfrac{1}{2}\dfrac{\partial v_r}{\partial z} + \dfrac{1}{2}\dfrac{\partial v_z}{\partial r} \\[2mm] 0 & \dfrac{v_r}{r} & 0 \\[2mm] \dfrac{1}{2}\dfrac{\partial v_r}{\partial z} + \dfrac{1}{2}\dfrac{\partial v_z}{\partial r} & 0 & \dfrac{\partial v_z}{\partial z} \end{pmatrix}$$

(38)

The second component of functional (35) is responding for friction. To compute friction power on the boundary S of area V a model given by the subsequent equation was used:

$$\dot{W}_t = \int_S m \frac{\sigma_p}{\sqrt{3}} \|\bar{v}\| \, dS$$

(39)

In equation (39) m is the so called friction factor which is usually experimentally selected and $\bar{v}$ is a relative velocity vector of metal and tool $\bar{v} = v - v_t$. In case of tensile test the samples are permanently fixed in jaws of a physical simulator and friction must not be taken into account. However, compression test requires sharing the friction power which is significant.

The model of sample velocity field

Clearly defined deformation field resulting from the optimal solution of functional (37) cannot be calculated without one of the conditions mentioned before. For the solid zones the incompressibility condition can be described by universal operator equation independently of the mechanical state of the deformation process:

$$\nabla v = 0$$

(40)

Because the semi-solid zone is characterized by density change due to still ongoing progress of steel state of aggregation, the condition of incompressibility is inadequate to reflect changes and was replaced with the mass conservation condition, which describes the following modified operational equation:

$$\nabla v - \frac{1}{\rho}\frac{\partial \rho}{\partial t} = 0$$

(41)

The basis for the optimization of functional (35) is the velocity field determined by appropriate system of velocity functions in the concerned area. These functions are then the source of deformation field and other physical quantities affecting the power functional formulation. Obtaining an accurate real velocity field requires the use of velocity functions depending on a number of variational parameters. The functions should be flexible enough to map the field throughout the whole volume of the deformation zone. Analytical description of each component of the velocity field with a single function in the whole area of deformation is not preferred. This approach creates difficulties especially in areas not subjected to the deformation where the velocity function should remain constant. Therefore, the solution to the problem of semi-solid metal flow was based on the method proposed by Malinowski in (Malinowski, 1986, 1997, 2005). This method involves the breakdown of the elements and the deformation velocity field approximation by polynomials with coefficients different for each element. The method was originally applied to solutions with a constant volume. The author of the current paper has developed a new method for semi-solid materials by adapting the source one to the analysis of materials with variable density. In the case of deformation of axial-symmetrical bodies the incompressibility condition is given by following differential equation:

$$\frac{\partial v_r}{\partial r} + \frac{v_r}{r} + \frac{\partial v_z}{\partial z} = 0$$

(42)

For the semi-solid area equation (42) is replaced by the mass conservation condition due to existing density changes. The longitudinal velocity has been calculated as an analytical function of radial velocity using this condition. In cylindrical coordinate system the condition has been described with an equation:

$$\frac{\partial v_r}{\partial r} + \frac{v_r}{r} + \frac{\partial v_z}{\partial z} - \frac{1}{\rho}\frac{\partial \rho}{\partial \tau} = 0$$

(43)

Equation (42) is a special case of equation (43) and therefore the proposed solution will consider the dependence (43) as more general. In (43) ρ is the temporary material density and τ is the time variable. The proposed variational formulation makes the longitudinal velocity dependent on the radial one. Condition (43) allows for the calculation of $\partial v_z/\partial z$ derivative as a function of $\partial v_r/\partial r$ after analytical differentiation of radial velocity distribution function $v_r(r,z)$. Hence, the longitudinal velocity is calculated as a result of analytical integration according to following equation:

$$v_z = - \int \left(\frac{\partial v_r}{\partial r} + \frac{v_r}{r} - \frac{1}{\rho}\frac{\partial \rho}{\partial \tau}\right) dz$$

(44)

In this case the velocity field depends only on one function – the radial velocity distribution. Both the components (V_r and V_z) satisfy the mass conservation imposed on the velocity field. The functional takes the form of equation (35) and in case of application of one of the methods requiring discretization (FEM, FDM or any meshless method) the number of discrete parameters is significantly reduced (at least by half). Only the right class of the velocity field distribution functions is problematic. The functions must be relevant for description of the material deformation and sufficiently flexible. Hence, the whole control volume is usually divided into sub-areas and the functions are defined in local coordinate systems for each sub-region. It requires the definition of both the local system and transformation from local to global one. The r coordinate acts as an independent variable (abscissa) in global area and varies in the range of r_m to R_m. The z coordinate depends on r and is limited by functions describing both the area boundaries: lower $z_l = f(r)$ and upper $z_u = g(r)$. Considering all the assumptions two linear functions, binding both the systems - global Orz and local one $O\xi\eta$ were defined

$$\xi(r,z) = \frac{r}{R_m}$$

$$\eta(r,z) = \frac{2z - g(r) - f(r)}{g(r) - f(r)}$$

(45)

The main assumption of the presented model is the dependence of the longitudinal velocity distribution function $v_z(\xi,\eta)$ on the radial velocity distribution function $v_r(\xi,\eta)$. For this purpose, the form of the function V_r has to be determined on the basis of analysis of the velocity field components distribution in the control area. For further discussion one assumes the following form V_r function:

$$v_r(\xi,\eta) = \frac{1}{2}\frac{rv_0}{g(r)-f(r)}\left(1+\frac{\partial\psi(\xi,\eta)}{\partial\eta}\right)$$

(46)

where v_0 is the GLEEBLE jaw velocity and $\psi(\xi,\eta)$ is a distribution function of velocity field components in local coordinate system. It should be remembered that for the areas in which the steel is in solid state the incompressibility condition given by dependence (42) should be taken into account and for zones with semi-liquid steel mass conservation equation (43) is valid. Linking the longitudinal velocity v_z with the radial one is done precisely through these two conditions. Taking into account the more general equation (43) and assuming a known value of the radial velocity one can be determine the longitudinal one using the following dependence:

$$v_z(\xi,\eta) = \int\left[\frac{1}{\rho}\frac{\partial\rho}{\partial t} - \frac{\partial v_r(\xi,\eta)}{\partial r} - \frac{v_r(\xi,\eta)}{r}\right]dz$$

(47)

The consequence of such a conduct is the fact that this condition is imposed on the velocity field in an analytical form. As already mentioned it is of major importance for optimizing the correct flow field for the steel being in semi-solid conditions.

In order to relate both the velocities the derivative of the velocity with respect to the radial coordinate has to be calculated first. Having in mind the dependence of f and g on r and similar one of ψ on ξ and on η one can write:

$$\frac{\partial v_r}{\partial r} = \frac{v_0}{2}\left[\frac{\partial}{\partial r}\left(\frac{r}{g-f}\right)\left(1+\frac{\partial\psi}{\partial\eta}\right) + \frac{r}{g-f}\frac{\partial}{\partial r}\left(1+\frac{\partial\psi}{\partial\eta}\right)\right]$$

(48)

After some differentiations and arrangements relationship (48) can be written in a form:

$$\frac{\partial v_r}{\partial r} = \frac{v_0}{2(g-f)}\left(1+\frac{\partial\psi}{\partial\eta}+\xi\frac{\partial^2\psi}{\partial\xi\,\partial\eta}\right) - \frac{rv_0\left(\frac{\partial g}{\partial r}-\frac{\partial f}{\partial r}\right)}{2(g-f)^2}\left[1+\frac{\partial\psi}{\partial\eta}+\frac{\partial^2\psi}{\partial\eta^2}\left(\frac{\frac{\partial g}{\partial r}+\frac{\partial f}{\partial r}}{\frac{\partial g}{\partial r}-\frac{\partial f}{\partial r}}+\eta\right)\right]$$

(49)

Taking into account equation (49) and relationship (43) one can calculate the derivative of the longitudinal velocity with respect to z.

$$\frac{\partial v_z}{\partial z} = -\frac{\partial v_r}{\partial r} - \frac{v_r}{r} + \frac{1}{\rho}\frac{\partial \rho}{\partial \tau} = \frac{r v_0 \left(\frac{\partial g}{\partial r} - \frac{\partial f}{\partial r}\right)}{2(g-f)^2}\left[1 + \frac{\partial \psi}{\partial \eta} + \frac{\partial^2 \psi}{\partial \eta^2}\left(\frac{\frac{\partial g}{\partial r} + \frac{\partial f}{\partial r}}{\frac{\partial g}{\partial r} - \frac{\partial f}{\partial r}} + \eta\right)\right] -$$

$$- \frac{v_0}{2(g-f)}\left(1 + \frac{\partial \psi}{\partial \eta} + \frac{1}{2}\xi\frac{\partial^2 \psi}{\partial \xi\,\partial \eta}\right) + \frac{1}{\rho}\frac{\partial \rho}{\partial \tau}$$

(50)

After appropriate integration the velocity is given by the following relationship:

$$v = -\frac{v_0}{4}\left\{2\left(\eta + \frac{g+f}{g-f} + \psi\right) + \xi\frac{\partial \psi}{\partial \xi} - \frac{\left(\frac{\partial g}{\partial r} - \frac{\partial f}{\partial r}\right)}{g-f}\left[\eta + (1-r)\psi + r\left(\frac{\frac{\partial g}{\partial r} + \frac{\partial f}{\partial r}}{\frac{\partial g}{\partial r} - \frac{\partial f}{\partial r}} + \eta\right)\frac{\partial \psi}{\partial \eta}\right]\right\} +$$

$$+ \frac{\eta(g-f) + g + f}{2\rho}\frac{\partial \rho}{\partial \tau}$$

(51)

Function $\psi = \psi(\xi, \eta)$ occurring in all the relationships describing the velocity field can be under the Weierstrass theorem approximated by polynomials. Approximation of $\psi(\xi, \eta)$ with the help of one polynomial in the whole deformation zone, although possible in some cases, is impractical and is a source of many problems. On the other hand the division of areas into smaller sub-areas requires continuity. To ensure continuity of the velocity field and strain field in the whole zone of deformation, including the boundaries of the subregions, function $\psi(\xi, \eta)$ should be at least of class C^2.

DENSITY CHANGES AND THEIR INFLUENCE ON REMAINING MODELS

In the proposed solution one of the most important parameters is the density. Its changes influence the mechanical part of the presented model and strongly depend on the temperature. The knowledge of effective density distribution is very important for modeling deformation of mushy materials. In the presented solution a model of density changes based on empirical data was applied.

Density distribution is one of the most important properties of the mushy steel which is subjected to the deformation. Its changes have influence on both the mechanical and thermal parts of the presented model. On the other hand, the density is strongly dependent on the temperature. Moreover, the solidification process causes non-uniform density distribution in the controlled volume. Since, the knowledge concerning effective density distribution is very important for the behavior of deformation of porous and mushy materials and the modeling of such species requires good density changes model.

Density variations of liquid, semi-solid and solid materials are ruled by three phenomena:

- solid phase formation,
- laminar liquid flow through porous material and
- thermal shrinkage.

Transient rate of density changes is ruled by an equation:

$$\frac{\partial \rho}{\partial \tau} = \frac{\partial \rho_p}{\partial \tau} + \frac{\partial \rho_f}{\partial \tau} + \frac{\partial \rho_t}{\partial \tau}$$

(52)

In (52) the subsequent right hand derivatives of ρ_p, ρ_f i ρ_t with respect to transient time variable τ denote the density changes as a result of three mentioned phenomena One may calculate the density changes due to solid phase formation according to the relationship:

$$\frac{\partial \rho_p}{\partial \tau} = [\rho_s(1 - X_l) + \rho_l X_l]\left(\frac{\rho_s}{\rho_l} - 1\right)\frac{\partial X_l}{\partial \tau}$$

(53)

where X_l and X_s are the shares of liquid and solid phases are semi-steel. Changes in density caused by laminar flow of the liquid phase through the porous material are described by the equation:

$$\frac{\partial \rho_f}{\partial \tau} = \rho_l X_l \left(\frac{\partial v_r}{\partial r} + \frac{v_r}{r} + \frac{\partial v_z}{\partial z}\right)$$

(54)

In (54) v is the velocity of the metal particles flow. Changes in density due to thermal shrinkage depend on the speed of changes in temperature and coefficients of linear thermal expansion β_s i β_l of both solid and liquid phases:

$$\frac{\partial \rho_t}{\partial \tau} = [\beta_s \rho_s(1 - X_l) + \beta_l \rho_l X_l]\frac{\partial T}{\partial \tau}$$

(55)

where T is the temperature on an absolute scale. Issues of density changes mechanisms were the subject of (Glowacki, 2002). Changes in the density as a result of the velocity and temperature of the metal particles substantially complicate the problem of optimizing the metal flow velocity field. Coupled solution of all the problems is difficult and very often an uncoupled model is used.

Empirical model of density changes

The density changes model is rather complex and its solution is associated with an additional increase in computational complexity of the total solution. Regardless of the solution used the development of a right model is a problem in itself. It requires addressing a number of issues related to the change of state, the flow of the liquid phase in the presence of solid steel frames, etc. This is an important issue - however, it requires commitment of substantial computer resources and long computation times. Hence another way of taking density into consideration is possible due to temperature dependency of this quantity (Glowacki, 1996). In order to avoid additional problems with solution of differential equation, density changes were calculated according to an empirical model taking into consideration experimental data. The model is slightly less accurate but such a method makes the solution much easier. The solution seems to be a good alternative way to predict changes in mushy steel. In proposed approach the density is depending on:

- temperature,
- chemical composition of the material and,
- steel microstructure

    The study published in (Glowacki, 1998), which is result of investigation carried out for steel in the solid state, shows that for typical forming processes impact of a steel grade on change in the density resulting from temperature changes is small. For determination of density in these conditions for both carbon and low-alloy steels it is proposed to apply following empirical equation:

$$\rho = \frac{7850}{(1 + \Delta l)^3}; \left[\frac{\text{kg}}{\text{m}^3}\right]$$

(56)

In equation (56) Δl is calculated according to following formula:

$$\Delta l = 0,004 \left(\frac{T + 273}{1000}\right)^2$$

Similar equation can be used for austenitic steels:

$$\rho = \frac{7897}{(1 + \Delta l)^3}; \left[\frac{\text{kg}}{\text{m}^3}\right]$$

(57)

The Δl parameter from (57) is calculated as:

$$\Delta l = -0{,}00358 + 0{,}00947\frac{T + 273}{1000} + 0{,}0103 \left(\frac{T + 273}{1000}\right)^2 - 0{,}00298 \left(\frac{T + 273}{1000}\right)^3$$

Similar dependence can be used for high-alloy steels. In this case it is necessary to modify the equation (56) in a manner appropriate for the particular steel grade. Thus, for temperature range which is proper for traditional process of steel hot deformation the calculation of changes in density seems to be pretty simple. Such temperatures are characteristic for certain sample areas.

Otherwise presents itself the problem for higher temperature ranges, where the deformation occurs during the simultaneous metal solidification. Here the density variations may be significant. For purposes of the current mathematical model an approach proposed by Mizukami was used (Mizukami at al., 2002). For carbon steels containing no other elements the density changes are functions of temperature. Steels were tested with a wide range of carbon content, which ranges from 0.005% to 0.56% by mass. The authors develop tests for typical steels having chemical composition expressed in% by mass given in Table. 1.

Left side of Figure 1 shows the change in density for MC1 grade steel as a function of temperature. For steel changing its states of aggregation some plots of density in the various phases of the transformation process has been developed. The right side of Figure 1 shows the course of the changes in the density of liquid phase as a function of ΔT_{l}– undercooling temperature with respect to the liquidus line. Changes in density are presented in relation to the base density of $7060 \; [kg/m^3]$.

Table 1: Chemical composition (mass %) of typical steels tested by authors of (Mizukami at al., 2002).

Steel	ULC	LC	MC1	MC2	HC
[C]	0.005	0.040	0.110	0.140	0.550
[Si]	0.010	0.040	0.100	0.160	0.150
[Mn]	0.120	0.190	0.480	0.540	0.910
[P]	0.014	0.026	0.020	0.016	0.021
[S]	0.003	0.006	0.008	0.003	0.001

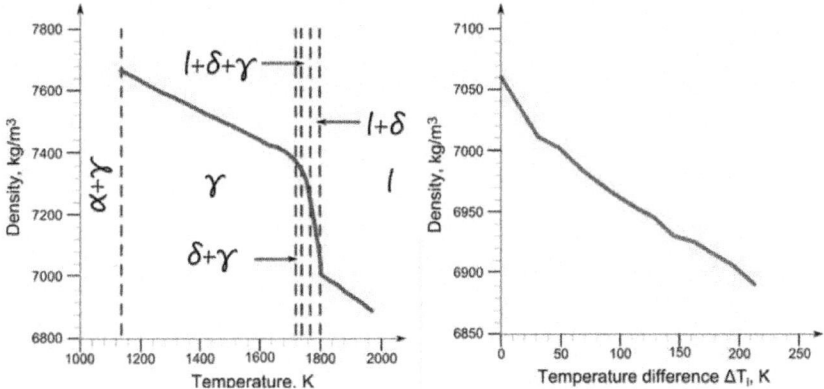

Figure 1: Density of MC1 grade steel as a function of temperature (left) and density changes of its liquid phase as a function of temperature increase (right). Plots are based on data published in (Mizukami at al., 2002).

Subsequent charts presented in Figure 2 show changes in density of δ i γ phases, respectively. Both of them are functions of undercooling temperature ΔT_δ and ΔT_γ of appropriate phases with respect to solidus temperature.

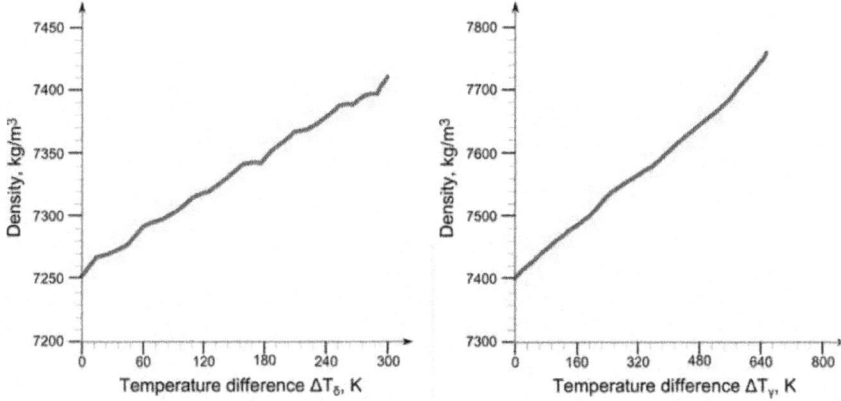

Figure 2: Density changes of steel phase δ (left) and γ (right) of MC1 steel grade as a function of temperature increase – based on data published in (Mizukami at al., 2002).

The presented graphs were used to develop analytical dependencies, which describe changes in the density of steel during the transformation of state of aggregation. In the region of coexistence of δ and γ phases density was estimated using the additivity rule. The correctness of this approximation was verified by comparing the theoretical results with those which were obtained from the measured values.

The density of carbon steels depends on the temperature and the existing fraction of liquid phase. The effect of alloying elements (except of coal) on the density of each steel phase is small, although the concentration of these components significantly affect the fraction of the phases. The density of each phase is calculated according to the following equations:

$$\rho_l = 7{,}02 - 5{,}50 \cdot 10^{-4}\,\Delta T_l$$
$$\rho_\delta = 7{,}27 + 3{,}07 \cdot 10^{-4}\,\Delta T_\delta$$
$$\rho_\gamma = 7{,}41 + 4{,}80 \cdot 10^{-4}\,\Delta T_\gamma$$

$$(58)$$

In equation (58) ρ_l, ρ_δ and ρ_γ indicate densities of liquid steel and its δ and γ phases, respectively. The density in the regions of occurrence of several phases simultaneously is given by the following equations:

$$\rho_{l+\delta} = \rho_l^0 + \Delta\rho_{l/\delta} \cdot X_\delta$$
$$\rho_{l+\gamma} = \rho_l^0 + \Delta\rho_{l/\gamma} \cdot X_\gamma$$
$$\rho_{l+\delta+\gamma} = \rho_l^0 + \rho_{l/\delta} \cdot f_\delta + \Delta\rho_{l/\gamma} \cdot f_\gamma$$

$$(59)$$

where ρ_l^0 denotes the density of the liquid phase for temperature discrepancy ΔT_l, $\Delta\rho_{l/\delta}$ and $\Delta\rho_{l/\gamma}$ are density differences between δ and γ phases for temperature drop from liquidus to solidus level, X_δ and X_γ are the fractions of δ i γ phases in surrounding liquid phase, respectively. and finally f_δ i f_γ are relative fractions of δ i γ phases. The density of $\delta+\gamma$ phase was estimated according to relationship:

$$\rho_{\delta+\gamma} = \rho_\delta \cdot X_\delta + \rho_\gamma \cdot X_\gamma$$

$$(60)$$

MUSHY STEEL FLOW STRESS CURVES DEVELOPMENT

The subsequent part of the chapter deals with the computation of mushy steel flow stress curves based on the developed mathematical model which helps to avoid interpretational problems occurring in traditional testing procedures.

Proper interpretation of the experimental results is possible with the help of appropriate computer aided testing system. Such a user friendly dedicated computer system with variable density has been developed (Glowacki & Hojny, 2009; Hojny & Glowacki, 2009a). The system codename called Def_Semi_ Solid is a result of theoretical research conducted in a team lead by the chapter author with the financial support of grants awarded by Polish Committee of Scientific Research. The system in itself is not a subject of the chapter and its details are not discussed. The program was developed using an object oriented technique and is compatible with both Windows and Unix based platforms.

During experiments a few quantities were recorded. Among them the most important are GLEEBLE jaws displacement, force and temperature. This is a start point for the inverse analysis. The system calculates the shape and size of the deformation zone and strain and stress fields as well as optimal values of flow stress curve parameters. The model described in the previous section allows for the comparison of theoretical and experimental results for non-uniform temperature field. Isothermal tests in the temperature range over 1400 °C are impossible even using sophisticated equipment like GLEEBLE simulator. The presented model is a solution to the experimental problems. The analysis of metal flow in subsequent regions of the sample deformation zone requires adequate methods. Classical techniques of interpretation of results of compression testing procedures fail due to significant samples barrelling which is inevitable at any temperature close to solidus level and which requires right analysis of metal flow in subsequent regions of the sample deformation zone.

A number of steel grades were subjected to series of experiments in Institute for Ferrous Metallurgy in Gliwice, Poland using GLEEBLE 3800 simulator. Example results of examination of two steels are reported in the current contribution. The first one is the 18G2A grade steel having 0.16% of carbon and the second was the S355J2G3So grade with 0.11% of carbon content. The essential aim of the investigation was the reconstruction of both temperature changes and strain evolution on specimen exposed to simultaneous deformation and solidification. The inverse procedure has been reported in (Glowacki & Hojny, 2009). Example results of inverse analysis are shortly described in succeeding subsections.

CHARACTERISTIC TEMPERATURE LEVELS

As mentioned before, apart from the liquidus and solidus temperatures, four other temperature levels are characteristic for the mushy steel behaviour. All the levels split the liquidus-solidus range into intervals. The most important for the extra high temperature rolling process design is the nil ductility temperature (NDT). The plastic deformation of a steel specimen is possible only below the

NDT temperature. The temperature levels have to be calculated according to results of series of difficult experiments which are not a subject of the current paper. For carbon steels with the carbon content of around 0.1 % the equilibrium liquidus and solidus temperature levels are 1523°C and 1482°C, respectively and the NDT is 1420°C. One must note that the last one is a conventional temperature of a sample surface (indicated during experimental procedure). The maximum and minimum temperatures in the sample's central cross-section may differ by 60- 70 °C. The equilibrium liquidus and solidus temperatures for 18G2A grade steel are 1513°C and 1465°C, respectively. The measured mean value of NDT temperature of the steel falls into the range of 1420°C÷1425°C. The NDT is related to the temperature at which the last liquid phase particles existing in the central part of the sample disappear in static processes. It has been observed that for temperatures higher than NDT a remainder of liquid phase still exist in the central part of the sample (Hojny & Glowacki, 2009a). For dynamic cooling and deformation processes in some regions of the sample the remainder of liquid phase can be observed at temperatures lower than NDT because the difference between sample surface and its central region is higher than for quasi-static processes.

Yield stress functions

The well-known Voce formula (Voce, 1955) was adopted for the description of the shape of yield stress function. Figure 3 presents four subsequent stages of an example compression test at higher sample surface temperature, i.e. 1425°C for the quasi-static process. One can observe that the experiment was successful (no metal outflow) and the deformation of the sample was realised despite the significant barrelling of the sample.

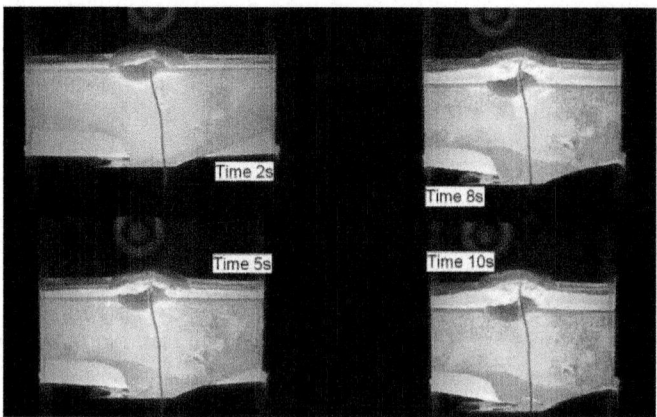

Figure 3: Four stages of the deformation process ran at temperature 1425°C for the quasi–static deformation process. The figure presents the central part of the sample.

Due to significant strain inhomogeneity inverse analysis is the only method allowing for appropriate calculation of coefficients of yield stress functions at any temperature higher than NDT. The objective function of the analysis was defined as a norm of discrepancies between calculated (f^c) and measured (F^m) loads in a number of subsequent stages of the compression according to the following equation:

$$\varphi(x) = \sum_{i=1}^{n}(F_i^c - F_i^m)^2$$

(61)

where n is the number of subsequent intervals of stress versus strain curve. The theoretical forces f^c were calculated with the help of sophisticated numerical solver being the implementation of the model which was described in this chapter. Due to the very low level of recorded stresses the experimental curves obtained from the GLEEBLE machine are noisy. Before the application of inverse analysis they were smoothed using Fast Fourier Transformation (FFT) algorithm. The final shape of the curves for 18G2A and S355J2G3So grade steels after interpretation using inverse analysis are presented in figures 8 and 16, respectively. Figure 4 summaries the results of calculation of example coefficients of Voce formula for 18G2A grade steel which was deformed in a quasi-static process. The effective strain inside the deformation zone varied from 0 to 0.6 and the effective strain rate reached its maximum value of 2.9 s-1 in final stage of the deformation process. The presented curves are plotted using the calculated coefficients of Voce curve for temperature levels observed in the sam ples' crosssections and for strain rate equal to 1 s^{-1}.

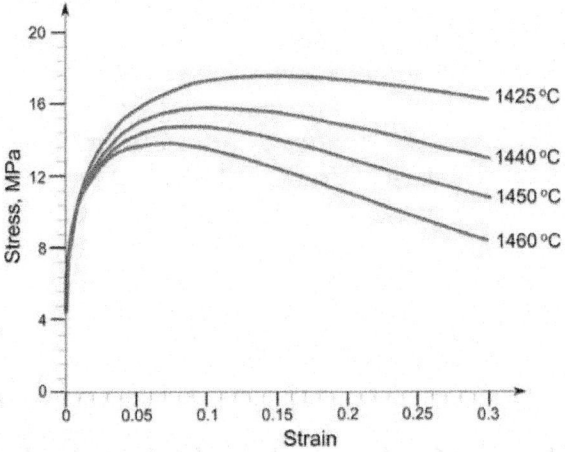

Figure 4: Flow stress vs. strain at several temperature levels for 18G2A steel grade deformed during quasi-static process – strain rate 1 s-1 (Glowacki & Hojny, 2010).

Example results of investigation of S355J2G3So grade steel are presented in Figure 5. The investigation procedures were analogous to those applied in case of18G2A grade steel.

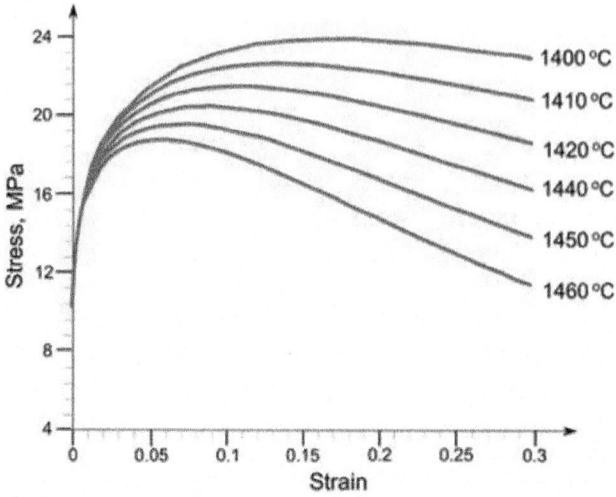

Figure 5: Stress-strain curves at several temperature levels from the range of 1400-1450°C (S355J2G3So grade steel, quasi-static process – strain rate 1 s-1 (Glowacki & Hojny, 2010).

CONCLUSIONS

Modeling of deformation of steel samples with mushy zone requires resolving several problems which are characteristic for the temperature range close to the solidus level. Some of the problems are independent of the strain and stress state of the material and are similar for both axial-symmetrical and three dimensional cases. The computation of characteristic temperatures and temperature-dependent sudden changes of steel plastic properties require advanced methods of computer simulation. The most important property for material plastic behavior is the yield stress function describing strain-stress curve. The proposed analytical model allows computation of such kind relationships.

The chapter has been dedicated to hybrid numerical-analytical model of semi-solid steel behaviour under plastic deformation. Application of inverse analysis and the proposed model allows for the testing of rheological properties of steels at temperature higher than 1400 °C. The results of the research are crucial for a unique computer system allowing for proper interpretation of the results of very high temperature compression tests. The classical interpretation of such results is improper due to strong strain inhomogeneity. The developed system is a tool to overcome many interpretational problems allowing for the computation of the appropriate shape and parameters of yield-stress curves. The curves have crucial influence on the results of computer simulation of semi-solid steel deformation.

The model presented in the current contribution is an axial-symmetrical one. The author have run further investigations leading to the development of a fully three-dimensional model of integrated casting and rolling processes as well as the soft reduction process, that is a part of strip casting technology. Like the model presented in the hereby chapter the spatial one also focuses on three main aspects: thermal, mechanical and density changes. Further intention of the research is the development of fully three dimensional model of mushy steel behaviour during rolling of plates with mushy region. The model will be useful for technologists working on the development of an integrated casting and rolling process. It is the most recent technology of sheet steel production, which is very profitable and requires extremely low energy consumption – very important for steel and automotive industries.

The compression tests carried out have shown good predictive ability of the proposed solution. They show that the flow stress above the NDT is strongly temperature and strain rate dependent. Low carbon steels, having carbon content of 0.11% and 0.16%, have been investigated in wide temperature range and strain rate. Example results of the experimental work were presented delivering a set of equations describing rheological behaviour of the investigated steels. The presented model and experimental procedure requires further investigation leading to the improvement of the solution and modelling additional phenomena accompanying the simultaneous deformation and solidification processes.

ACKNOWLEDGMENTS

The work has been supported by the Polish Ministry of Science and Higher Education - Grant No. N N508 585539.

REFERENCES

1. Adhikari S.K. (1998). Variational principles for the numerical solution of scattering problems. Wiley, New York USA, ISBN 0471181935

2. Bower, A.F. (2010) Applied mechanics and solids, CRC Press – Taylor & Francis Group, New York USA, ISBN 987-1-4398-0247-2

3. Chakrabarty, J. (2006). Theory of plasticity. Elsevier Butterworth-Heinemann, Oxford UK, ISBN 978-0-7506-6638-2

4. Evans, L.C. (1998), Partial Differential Equations, American Mathematical Society, ISBN 0821807722.

5. Findaeisen, W., Szymanowski, J., Wierzbicki, A (1980). Theory and optimization methods. PWN, Warszawa, ISBN 8301009764

6. Glowacki, M. (1996). Finite element three-dimensional modelling of the solidification of a metal forming charge, Journal of Materials Processing Technology, Vol. 60, No. 1-4, pp. 501-504, ISSN 0924-0136

7. Glowacki, M. (1998). Thermal-mechanical–microstruktural model of shape rolling. Dissertations and Monographs, Vol. 76, AGH Publishing, Krakow Poland, ISSN 0867-6631

8. Głowacki, M. (2002) Possibilities of mathematical modeling of deformation of samples with mushy zone, Proceedings of 44th Mechanical Working and Steel Processing Conference, pp. 1151-1162, Orlando USA, September 1, 2002, ISBN 978-1-886362-62-8

9. Glowacki, M. (2006). Mathematical modelling of deformation of steel samples with mushy zone, In: Research in Polish metallurgy at the beginning of XXI century, Committee of Metallurgy of the Polish Academy of Science, K. Swiatkowski, (Ed.), 305-324,

10. Publishing House Akapit, Krakow Poland Glowacki, M. & Hojny, M. (2006). Development of a computer system for high temperature steel deformation testing procedure, Proceedings of Simulation, Design and Control of Foundry Processes, pp. 145-156, Krakow Poland, November 22-24, 2006

11. Glowacki, M. & Hojny, M. (2009). Inverse analysis applied for determination of strain-stress curves for steel deformed in semi-solid state, Inverse Problems in Science and Engineering, Vol.17, No. 2, pp. 159–174, ISSN 1741-5977

12. Glowacki, M. & Hojny, M. (2010). Investigation of mushy steel rheological properties at temperatures close to solidus level, In: Polish metallurgy 2006–2010 in time of the worldwide economic crisis, Committee of Metallurgy of the Polish Academy of Science, K.

13. Swiatkowski, (Ed.), 193-212, Publishing House Akapit, Krakow, Poland, ISBN 978-83-60958-59-9

14. Glowacki, M., Hojny, M. & Jędrzejczyk, D. (2010). Hybrid analytical-numerical system of mushy steel deformation. In : Recent studeis in meshless & other novel computational methods, B. Sarler & S.N. Atluri, (Eds.), pp. 35-54, Tech Science Press, ISBN-10 0-9824205-4-4, USA

15. Hojny, M. & Glowacki, M. (2008). Computer modelling of deformation of steel samples with mushy zone, Steel Research International, vol. 79, No. 11, (2008), pp. 868-874, ISSN 1611-3683

16. Hojny, M. & Glowacki, M. (2009a) The methodology of strain – stress curves determination for steel in semi-solid state, Archives of Metallurgy and Materials, Vol. 54, No. 2, pp. 475–483, ISSN 1733-3490

17. Hojny, M. & Glowacki, (2009b) The physical and computer modelling of plastic deformation of low carbon steel in semi-solid state, Transactions of the ASME, Journal of Engineering Materials and Technology, Vol. 131 No. 4, pp. 041003-1–041003-7, ISSN 0094-4289

18. Hojny, M., Glowacki, M. & Malinowski Z. (2009), Computer aided methodology of strainstress curve construction for steels deformed at extra high temperature, High Temperature Materials and Processes, Vol. 28, No. 4, pp. 245–252, ISSN 0334-6455

19. Hojny, M. & Glowacki, M. (2011). Modeling of Strain-Stress Relationship for Carbon Steel Deformed at Temperature Exceeding Hot Rolling Range, Transactions of the ASME, Journal of Engineering Materials and Technology, Vol. 133, No. 2, pp. 021008-1–021008-7, ISSN 0094-4289

20. Kang, C.G. & Yoon, J.H. (1997). A finite-element analysis on the upsetting process of semisolid aluminum material, Journal of Materials Processing Technology, Vol. 66, No. 1-3, pp. 76-84, ISSN 0924-0136

21. Koc, M., Vazquez, V., Witulski, T. & Altan, T. (1996). Application of the finite element method to predict material flow and defects in the semi-solid forging of A356

22. aluminum alloys, Journal of Materials Processing Technology, Vol. 59, No. 4, pp. 106- 112, ISSN 0924-0136

23. Kopp, R., Choi, J. & Neudenberger D. (2003). Simple compression test and simulation of an Sn–15% Pb alloy in the semi-solid state, Journal of Materials Processing Technology, Vol. 135, No. 2-3, pp. 317-323, ISSN 0924-0136

24. Leader, J. J. (2004). Numerical Analysis and Scientific Computation. Addison Wesley, Boston Massachusetts, ISBN 978-0-201-73499-7

25. Li, J.Y., Sugiyama, S. & Yanagimoto, J. (2005), Microstructural evolution and flow stress of semi-solid type 304 stainless steel, Journal of Materials Processing Technology, Vol. 161, No. 3, pp. 396-406, ISSN 0924-0136

26. Malinowski, Z. (1986). Analysis of upsetting process based on velocity fields, PhD thesis, AGH Krakow Poland, in Polish

27. Malinowski, Z. (1997). Effect of heat generation on flow stress deformation based on the axially symmetric compression test. Metallurgy & Foundry Engineering, 23, 1997, 459-467, ISSN 1239-2325

28. Malinowski Z. (2005). Numerical models in metal forming and heat transfer. Wyd. AGH Publishing, Krakow Poland, in Polish, ISBN: 83-89388-98-7

29. Mizukami, H., Yamanaka, A. & Watanabe, T. (2002). Prediction of density of carbon steels, ISIJ International, Vol. 42, No. 4, pp. 375-384, ISSN 0915-1559

30. Nocedal J. & Wright S.J. (2006). Numerical Optimization. Springer-Verlag, Berlin Germany. ISBN 0-387-30303-0

31. Pinchover, Y. & Rubinstein, J. (2005). An Introduction to Partial Differential Equations, New York: Cambridge University Press, ISBN 0521848865.

32. Polyanin, A.D. & Zaitsev, V.F. (2004). Handbook of Nonlinear Partial Differential Equations, Boca Raton: Chapman & Hall/CRC Press, ISBN 1584883553.

33. Sang-Yong, L., Jung-Hwan L. & Young-Seon L. (2001). Characterization of Al 7075 alloys after cold working and heating in the semi-solid temperature range. Journal of Materials Processing Technology, Vol. 111, No. 1-3, pp. 42-47, ISSN 0924-0136

34. Seol, D.J., Won, Y.M., Yeo, T., Oh, K.H., Park, J.K. & Yim, C.H. (1999). High Temperature Deformation Behavior of Carbon Steel in the Austenite and δ-Ferrite Regions, ISIJ International, Vol. 39, No. 1, pp. 91-98, ISSN 0915-1559

35. Seol, D.J., Oh, K.H., Cho, J.W., Lee, J.E., Yoon, U.S. (2002). Phase-field modelling of the thermo-mechanical properties of carbon steels, Acta Materialia, Vol. 50, No. 9, pp. 2259-2268, ISSN 1359-6454

36. Senk, D., Hagemann, F., Hammer, B., Kopp, R., Schmitz, H.P. & Schmitz, W. (2000). Umformen und Kühlen von direktgegossenem, Stahlband, Stahl und Eisen, Vol. 120, No. 6, pp. 65-69, ISSN 0340-4803

37. Suzuki, H.G., Nishimura, S. & Yamaguchi S. (1988). Physical simulation of the continuous casting of steels, Proceedings of Physical Simulation of Welding, Hot Forming and Con- tinuous Casting, pp. 166-191, Canmet Canada, May 2-4, 1988 Voce, E. (1955). A Practical Strain Hardening Function, Metallurgia, vol. 51, 1955, pp. 219-226, ISSN 0141-8602

38. Zhao Y.Q., Wu W.L. & Chang H. (2006). Research on microstructure and mechanical properties of a new α + Ti2Cu alloy after semi-solid

deformation, Materials Science and Engineering, Vol. 416, No. 1-2, pp. 181-186, ISSN 0921-5093

39. Zienkiewicz, O.C., Taylor, R. L. & Zhu, J.Z. (2005). The Finite Element Method: Its Basis and Fundamentals, Elsevier Butterworth-Heinemann, Oxford UK, ISBN 0-7506-6320-0

Chapter 5

OPTIMIZATION OF CAPACITIVE ACOUSTIC RESONANT SENSOR USING NUMERICAL SIMULATION AND DESIGN OF EXPERIMENT

Rubaiyet Iftekharul Haque[1,2,], Christophe Loussert[2], Michelle Sergent[3], Patrick Benaben[1] and Xavier Boddaert[1]

[1]Centre Microélectronique de Provence, Ecole des Mines de Saint-Etienne, Gardanne 13541, France

[2]TAGSYS RFID, 13600 La Ciotat, France

[3]Aix-Marseille Université, LISA EA 4672, 13397 Marseille Cedex 20, France

ABSTRACT

Optimization of the acoustic resonant sensor requires a clear understanding of how the output responses of the sensor are affected by the variation of different factors. During this work, output responses of a capacitive acoustic transducer, such as membrane displacement, quality factor, and capacitance variation, are considered to evaluate the sensor design. The six device parameters taken into consideration are membrane radius, backplate radius, cavity height, air gap, membrane tension, and membrane thickness. The effects of factors on the output responses of the transducer are investigated using an integrated methodology that combines numerical simulation and design of experiments (DOE). A series of numerical experiments are conducted to obtain output responses for different combinations of device parameters using finite element methods (FEM). Response surface method is used to identify the significant factors and to develop the empirical models for the output responses. Finally, these results are utilized to calculate the optimum device parameters using multi-criteria optimization with desirability function. Thereafter, the validating experiments are designed and deployed using the numerical simulation to crosscheck the responses.

INTRODUCTION

For many years, acoustic sensors have been used in many civilian and military applications, such as in cellular phones, hearing aids, and computers, in addition to high quality studio microphones for sound recording [1], sonar for underwater objects detection [2], and in the acoustic sensor systems for target acquisition and surveillance purposes [3] etc. An acoustic transducer provides analog output that is proportional to the variation of acoustic pressure acting upon a flexible diaphragm. Most familiar examples of acoustic sensors are the microphone, earphone etc. There are different types of acoustic sensors: namely, piezoelectric, piezoresistive, and capacitive [4]. Among them, capacitive acoustic sensors show the highest sensitivity while maintaining low power consumption [5]. Capacitive sensing is independent of the base materials and relies on the variation of the capacitance when geometry of a capacitor is changing. Furthermore, capacitive acoustic sensor can be used as both an active and passive sensing device.

A capacitive acoustic transducer is an electromechanical-acoustic system. It usually consists of a fixed backplate electrode and a flexible diaphragm that acts as a second electrode, separated by a dielectric material, such as air, to form a parallel plate capacitor. The deflection of the diaphragm occurs due to incident acoustic pressure, thereby providing capacitance variation in response to the change in air gap. In general, capacitive acoustic transducer suffers from over-damping, as a thin layer of air is trapped in between the electrodes; therefore, capacitive acoustic transducers are usually designed and fabricated with porous membranes or/and perforated backplates to reduce the damping effect.

To date, many capacitive acoustic sensors have been developed, and some of them are commercially available. However, its design varies based on the application domains, and the majority of these are targeted for audio applications with nearly uniform sensitivity over a relatively wide range of frequencies in the human hearing range, 20 Hz–20 kHz [4,5,6,7,8,9,10].

Recently, a new simplified design concept has been proposed to fabricate a capacitive acoustic transducer, which consists of a central cylindrical rigid backing electrode of small radius surrounded by a flat annular cavity below a vibrating membrane clamped at its periphery separated by an air gap, which provides good sensitivity and a large frequency bandwidth [9,10]. Honzik et al. [9] have reported that this design leads to a higher sensitivity, as well as a larger frequency bandwidth.

A capacitive acoustic sensor, similar to that of a condenser microphone, can also be used as an acoustic resonant sensor by modifying different parameters related to the device fabrication. The characteristics of the damping material and other geometric parameters determine the transducer bandwidth. Generally, transducers respond to incident acoustic pressure over the entire range of relevant frequencies, whereas resonant transducers provide higher sensitivity at their natural frequencies.

The design presented in Figure 1 can be a potential candidate to fabricate a capacitive acoustic resonant sensor with good selectivity at a certain frequency. During this work, we investigate the possibilities to develop the acoustic resonator based on this simplified design concept. To fulfill the specific system requirement, a capacitive resonant sensor with strong sensitivity at specified frequency with narrow bandwidth is desired. To do so, one needs to optimize structural parameters, such as membrane radius, backplate radius, air gap, cavity height, membrane thickness, of the design of the acoustic sensor (Figure 1). In addition, the membrane tension and material uses to fabricate the device needed to be optimized, as well.

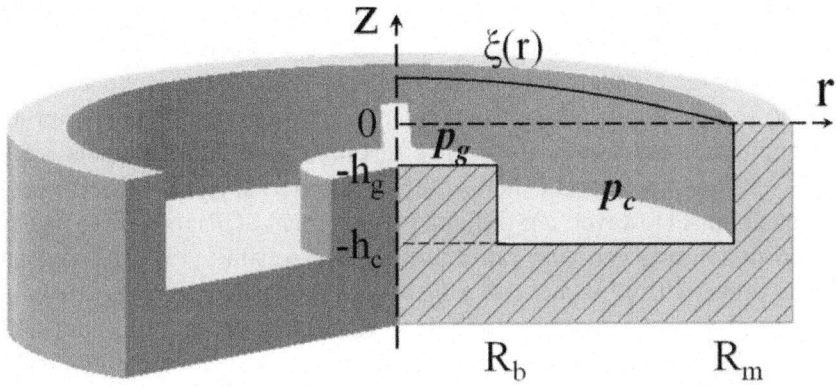

Figure 1: Schematic diagram of acoustic transducer.

As a large number of parameters are involved in acoustic sensor optimization, numerical simulation can be a powerful and economical tool for virtual device prototyping. However, the extensive computational effort is involved in numerical simulation and thus it usually takes a substantial amount of time to complete simulation runs of a complex structure. This paper presents a new design scheme for acoustic sensor optimization that combines numerical simulation using the COMSOL Multiphysics software and design of experiments (DOE) approach to optimize the acoustic sensor of the proposed design to obtain the acoustic resonator. DOE helps to develop a plan of experiments that provides a great deal of information about the effect of input parameters on responses. In this scheme, a set of numerical experiments is conducted to generate responses. Thereafter, based on the numerical simulation results, namely, membrane displacement, capacitance variation, quality factor, etc., the response surface method (RSM) is used to derive empirical models for each of the responses, which will later be used for optimization process. The empirical model reduces computational efforts in the acoustic sensor optimization, since they are far less complex than the original finite element model.

In case of a single response characteristic, optimization can simply be obtained by determining the experimental conditions that satisfied the expected response [11]. However, the performance of a capacitive acoustic resonant sensor is often characterized by a group of responses, such as static capacitance, capacitance variation, quality factor, etc. If more than one response comes into consideration, it is very difficult to select the optimal setting that can achieve all quality requirements simultaneously.

Vogel et al. [12] have applied FEM and a sequential quadratic programming (SQP) method as part of the CAPA optimization module to optimize micromachined capacitive ultrasound transducer array (CMUT), the design of comb structures for use in acceleration sensors, and the optimization of an electrostatic membrane device for an integrated silicone microphone. The SQP method is generally used for a nonlinearly constrained optimization problem that approximately solves a sequence of optimization subproblems, each of which optimizes a quadratic model of the objective subject to a linearization of the constraints. However, it is difficult to implement SQP methods so that exact second derivatives can be used efficiently and reliably [13]. The alternative of this approach, is to make use of a desirability function that transforms an estimated response into a scale-free value, known as global desirability [11,14,15]. With the multi-objective nature of our problem, desirability function is employed during this work to avoid the disadvantages of other methods.

The objectives of this study are to investigate the effect of different parameters of the transducer on the output responses using the numerical analysis and DOE approach, and to optimize device parameters to develop acoustic resonator that provides good sensitivity and selectivity. In this regard, the first part of this paper is devoted to the theoretical analysis to understand the system, and then the construction of the finite element (FE) model of the acoustic sensor (based on the design presented in Figure 1). Thereafter, DOE is introduced to achieve greater information about the effects of different input parameters on output responses with the least possible number of experiments. Finally, multi-criteria optimization is performed to obtain the optimum set of parameters, which is verified using numerical simulation.

THEORETICAL ANALYSIS

A capacitive acoustic sensor is an electro-mechanical transducer that transforms the mechanical deformation of the diaphragm in an output signal. The capacitance (C_0) of a parallel plate capacitor, with a fixed distance between the two electrodes h0 and area of overlap of the two electrodes plates Se, also known as effective area, is given by

$$C_0 = \frac{\varepsilon_0 \varepsilon_r S_e}{h_0}$$

(1)

where ε_0 represents electric constant (ε_0=8.854×10⁻¹²Fm⁻¹) and ε_r represents the relative static permittivity of the materials between the plates (for a vacuum, ε_r=1). When an external DC voltage (V_0) is applied, an electrostatic force (Fes) as presented by Equation (2), is created across the electrodes and induces a membrane deformation.

$$F_{es} = \frac{Q^2}{2\varepsilon_0 S_e} = \frac{\varepsilon_0 S_e}{2h_0^2} V_0^2$$

(2)

Thus the air gap (hg) becomes $(h_0 + \langle\xi\rangle_{es})$, where $\langle\xi\rangle_{es}$ represents the quiescent average deformation of membrane due to the electrostatic forces of the pre-polarization of the transducer. Therefore, the static capacitance (C_i) of the acoustic sensor can be expressed as,

$$C_i = \frac{\varepsilon_0 \varepsilon_r S_e}{h_0 + \langle\xi\rangle_{es}} = \frac{\varepsilon_0 \varepsilon_r S_e}{h_g}$$

(3)

The air gap of the transducer varies due to membrane deformation. If $\langle\xi\rangle_{Se}$ represents the average small-signal deformation; the varied distance between the back electrode and the membrane becomes $h_g + \langle\xi\rangle_{Se}$. Thus, the output capacitance (Cn) due to incident pressure can be expressed as follows,

$$C_n = \frac{\varepsilon_0 \varepsilon_r S_e}{h_g + \langle\xi\rangle_{Se}} = \frac{\varepsilon_0 \varepsilon_r S_e}{h_g}\left(1 - \frac{\langle\xi\rangle_{Se}}{h_g}\right) = C_i\left(1 - \frac{\langle\xi\rangle_{Se}}{h_g}\right)$$

(4)

where the expression has been expanded to the first order (Taylor series). Therefore, the capacitance variation (ΔC) can be obtained by subtracting Equation (4) from Equation (3),

$$\Delta C = |C_n - C_i| = C_i \frac{|\langle\xi\rangle_{Se}|}{h_g}$$

(5)

On the other hand, the total voltage (V) across the capacitor is the sum of the quiescent polarization voltage (V_0) and the small-signal output voltage (V_{out}). The charge (Q) in the capacitor can be expressed as,

Q=CV　　　　　　　　　　　　　　　　　　　　　　　　(6)

Its differentiation is given as:

dQ=CdV+VdC　　　　　　　　(7)

We assume that the system has a constant total charge, $Q=\sum i q i=$ const., thus dQ=0. The inclusion of this assumption in Equation (7) gives

$$dV = -\frac{dC}{C}V$$

(8)

Assuming $C=C_n$, $dC \approx \Delta C$, $dV \approx V_{out}$, $V \approx V0$, and introducing them into Equation (8):

$$V_{out} = V_0 \frac{C_i \frac{\langle\xi\rangle_{Se}}{h_g}}{C_i \left(1 - \frac{\langle\xi\rangle_{Se}}{h_g}\right)} = V_0 \frac{\langle\xi\rangle_{Se}}{h_g}\left(1 + \frac{\langle\xi\rangle_{Se}}{h_g}\right) \approx V_0 \frac{\langle\xi\rangle_{Se}}{h_g}$$

(9)

The expression has been expanded to the first order (Taylor series), and the higher order term is negligible and thus removed. Based on the analysis, it has been observed that the capacitance variation as well as output voltage of the acoustic sensor mainly depends on the membrane displacement. Therefore, to improve the sensitivity of the sensor, we have to design the sensor which will provide high membrane displacement.

Equations Governing the Membrane Displacement

The equation governing the vibration of the thin circular membrane of thickness tm, radius R_m, and density ρ_m under constant radial force per unit length (T_m) acting on its edge, driven by uniform harmonic incident acoustic pressure p_i over the membrane surface, loaded by the pressure field p(r), also known as reaction pressure at the membrane surface, can be expressed as [9,10,16,17]:

$$T_m\left(\Delta_r + K^2\right)\xi(r) = p_i - p(r), \quad 0 < r < R_m$$

(10)

Here, $\xi(r)$ being the vertical membrane displacement, Δr (equals to ∇^2_r) represents the Laplace operator, and K^2 defines the wavenumber of the free flexural vibration of the membrane,

$$K^2 = \frac{\omega^2 \rho_{ms}}{T_m} = \frac{\omega^2}{c^2}; \quad c = \sqrt{\frac{T_m}{\rho_{ms}}} \quad \text{and} \quad \rho_{ms} = t_m \rho_m$$

(11)

where, c denotes the speed of sound in the membrane, ρ_{ms} being the surface density or mass per unit area of the membrane and ω is the angular frequency. The membrane is supported on a rigid circular frame at its periphery $r=R_m$ (Dirichlet boundary condition), therefore

$$\xi(R_m)=0 \tag{12}$$

The reaction pressure p(r), loading the diaphragm, is due to the underlying air layer squeezed in the air gap and in the annular cavity under the membrane, where

$$p(r)=\begin{cases} p_g(r) & r\in(0,R_e) \\ p_c = const & r\in(R_e,R_m) \end{cases} \tag{13}$$

Here, pg(r) and pc(r) represent the pressure in the air gap and the pressure in the cavity volume which is assumed to be quasi-uniform, respectively, and R_e represents the effective radius and is equal to the radii of the back plate electrode (R_b).

The incident acoustic signal (with the time factor given by $e^{j\omega t}$) triggers the membrane displacement $\xi(r)$, which is assumed to be small and harmonic ($\xi(r)e^{j\omega t}$). The membrane displacement gives rise to the motion of the air in the domain below the circular membrane, composed of the air gap and annular cavity. It is assumed that the pressure variation in air gap and cavity region is constant throughout the thickness of the fluid film; it depends only on the tangent coordinate r. As the pressure variation does not depend on the z-coordinate, the z-component of the particle velocity (v) can be neglected. On the other hand, the temperature variation (τ), depends on both coordinates r and z, which is approximately proportional to the pressure variation outside the boundary layers. The temperature variation vanishes at the interfaces between the fluid layer and the membrane z=0, and between the fluid layer and the backing electrode z=−hg. Thus the boundary conditions associated with the system are,

$$v_{r(g,c)}(r,0)=v_{r(g,c)}(r,-h_{g,c})=0 \text{ and } \tau_{g,c}(r,0)=\tau_{g,c}(r,-h_{g,c})=0$$

(14)

The solution of the mean displacement of the circular membrane over the backplate electrode driven by the constant incident pressure pi due to the sound field can be expressed as follows [9,10]:

$$\langle \xi \rangle_{Se} = \frac{1}{S_e} \iint_{S_e} \xi(r) dS_e = \frac{2}{\sqrt{\pi} R_e R_m} \sum_n \xi_n \frac{J_1(K_n R_e)}{K_n J_1(K_n R_m)}$$

(15)

Pressure Sensitivity

The sensitivity level (L) of the acoustic sensor for the given polarization voltage V_0, represents the relation between the input pressure and the output voltage, and can be expressed as follows:

$$L = 20 \log \left[\left| \frac{V_{out}}{p_i} \right| \right] = 20 \log \left[\left| \frac{V_0 \langle \xi \rangle_{Se}}{p_i h_g} \right| \right]$$

(16)

Resonance Frequency

The selectivity of the acoustic resonant sensor depends on its natural frequency or resonance frequency. At resonance frequency, all parts of the membrane vibrate sinusoidally with the same frequency and with a fixed phase relation, which provides maximum displacement of membrane and is known as normal mode of vibration. Resonance frequencies of the membrane in vacuum are solely determined by its physical dimensions and mechanical constants: namely, Young's modulus, density of the membrane materials, size of membrane, and boundary conditions. As the maximum membrane displacement occurs at resonance frequency, it leads to the maximum sensitivity for the capacitive acoustic sensor. The natural frequencies of the pre-tensioned circular vibrating membrane in vacuum is given by [18,19],

$$f_{ij} = \frac{k_{ij}}{2\pi R_m} \sqrt{\frac{T_m}{t_m \rho_m}} = \frac{k_{ij}}{2\pi R_m} \sqrt{\frac{T_m}{\rho_{ms}}}$$

(17)

The values kij are derived from the roots of the Bessel functions of the first kind. The natural frequencies of vibration and mode shapes are identified by two integers (i, j) that characterize the mode shape. The index i=1, 2, 3, ... corresponds to the number of circumferential lines (with r=const.) on the membrane that have zero displacement, while j= 0, 1, 2, ... corresponds to the number of diametral lines (with θ=const.) that have zero displacement. The values kij for the first six modes are listed in Table 1.

However, in the case of a capacitive acoustic resonator, the membrane is usually loaded with an air cavity rather than vibrating in free space [20]. The presence of the cavity generally detunes the membrane resonance [21]. The shifts of first resonance frequency of the system towards the higher frequency than that of the membrane in vacuum occurs due to viscous damping and acoustic stiffness of the cavity.

Table 1: Values of kij derived from the roots of the Bessel functions of the first kind for first six modes.

Mode Number	Factor
1	$k_{10} = 2.4048$
2	$k_{11} = 3.8317$
3	$k_{11} = 3.8317$
4	$k_{12} = 5.1356$
5	$k_{12} = 5.1356$
6	$k_{20} = 5.5201$

Quality Factor

The quality factor (Qf), which is related to the energy loss of the vibrating diaphragm [22], is characterized by a resonator's bandwidth relative to its center frequency. Generally, in frequency domain, Q-factor is expressed as,

$$Q_f = \frac{f_r}{\Delta f} = \frac{\omega_r}{\Delta \omega}$$

(18)

where, f_r is the resonance frequency, Δf is the half-power bandwidth (i.e., the bandwidth over which the power of the vibration is greater than half the power at the resonance frequency), $\omega_r = 2\pi f_r$ is the angular resonance frequency and Δ_ω the angular half power bandwidth.

High Q_f value represents low damping, which indicate low rate of energy loss relative to the stored energy of the resonator [23,24]. Q-factor is inversely proportional to the damping coefficient of the oscillating system and define as [23],

$$Q_f = 2\pi \frac{W}{\Delta W} = \frac{m\omega_r}{\gamma} = \frac{K}{\gamma \omega_r}$$

(19)

where W is the total energy stored in the resonator, ΔW is the sum of the energy loss per cycle, K is the spring constant of the resonator, γ is the coefficient of the damping force, and m is the mass of the oscillator.

Thus, in the case of the acoustic resonator, higher Q-factor represents high selectivity. The Q-factor of the system can be improved by enhancing the total stored energy, while reducing the energy loss per cycle.

NUMERICAL SIMULATION

Finite Element Model (FEM)

The capacitive acoustic sensor works by transforming the mechanical deformation of the thin membrane (diaphragm), induced by an external incident pressure, into an AC voltage signal. Numerical simulation is performed using finite element method (FEM) not only to understand but also to quantify the effect of different input parameters on the membrane displacement, capacitance variation, Q-factor etc. The finite element simulation of the acoustic sensor is a moving boundary problem, in which the computational air domain within the sensor changes continuously, because of membrane vibration under harmonic acoustic wave. Three-dimensional (3D) FEM model is developed using half-slice of the air domain (symmetrical part), as illustrated in Figure 2, to reduce the computational time.

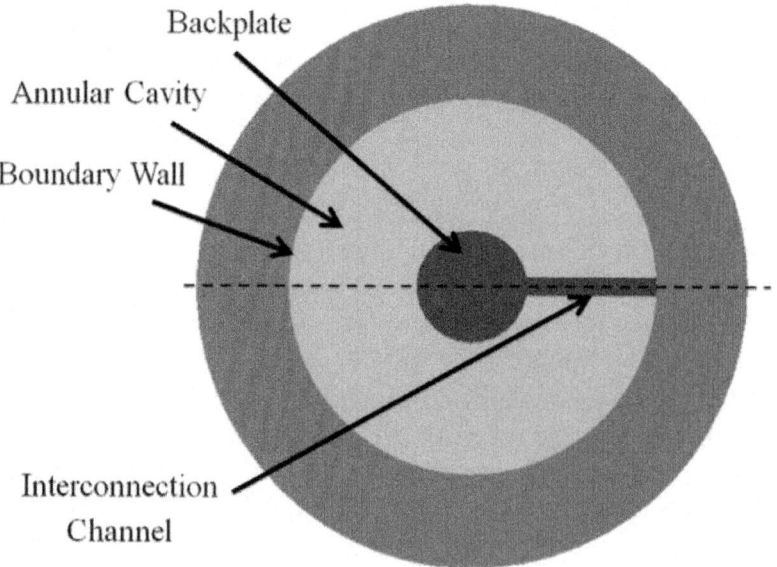

Figure 2: Schematic diagram of top view of the proposed acoustic sensor after removing the diaphragm.

Figure 3 illustrates the half-slice of the 3D air domain with finite element mesh, which was solved considering the periodicity and symmetry of the boundary value. The custom mesh is used in such a way that it resolves the acoustic boundary layer for the frequency range of 0 Hz to 250 kHz without mesh regeneration. The physical parameters of the air and the membrane materials are given in Table 2.

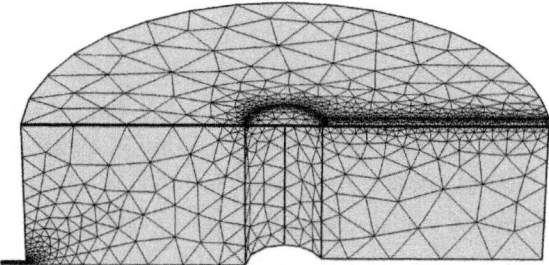

Figure 3: Half-slice of air domain (symmetrical part) of the acoustic sensor with finite element mesh.

Table 2: List of parameters of the thermoviscous fluid (air) and material properties of the membrane

Parameter	Value	Unit
Bulk viscosity (μ_{B0})	10×10^{-6}	Pa·s
Gas constant (R_{s0})	281.4	J/(kg·K)
Density of Membrane (ρ_m)	1390	kg/m^3
Young's modulus of membrane (E_m)	4×10^9	Pa
Poisson's ratio of membrane (υ_m)	0.38	-

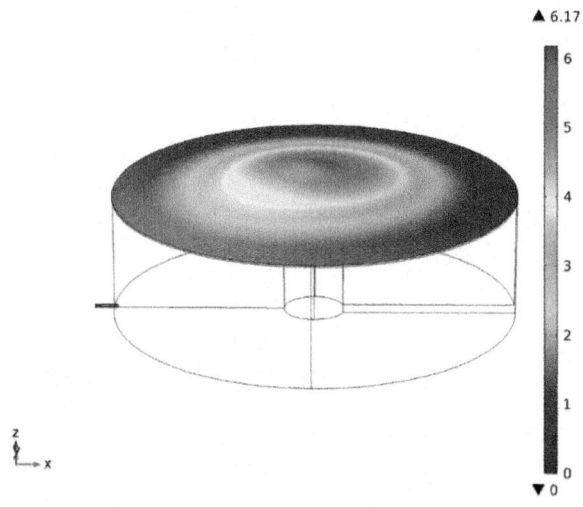

Figure 4: Measured displacement field of the membrane of the acoustic sensor at first resonance frequency ($f_r = 16{,}737$ Hz) for $R_m = 5$ mm, $R_b = 0.75$ mm, $h_c = 3000$ μm, $h_g = 30$ μm, $T_m = 500$ N/m and tm = 8 μm.

The resulting finite element model with fully coupled thermoacoustic, electrostatic, moving mesh and membrane physics interface was solved using the linear-perturbation solver, PARDISO, in the frequency domain. Solution of the numerical simulation provides the membrane displacement with respect to frequencies. Figure 4 presents the membrane displacement at first resonance frequency (fr).

COMSOL Multiphysics software (version 4.4) is used to perform 3D numerical simulation. All the numerical works have been executed on a workstation, DELL PRECISION T7600, having 32 Gigabytes RAM and 16 cores (two 3.1 GHz eight-core Intel Xeon E5-2687W processors).

The validation of the numerical model is checked by comparing the maximum membrane displacement of the numerical analysis with that of the theoretical analysis as presented in Equation (15). The results show proximate similarity as depicted in Figure 5. A little shift of the resonance frequency and a slightly smaller magnitude of the membrane displacement are obtained in the FEM results. They are caused by the presence of the interconnection channel to electrically connect the bottom electrode with the outside and venting hole in the geometry as it is in the real device; whereas for simplicity the effect of the interconnecting channel and venting hole are not considered in the theoretical analysis.

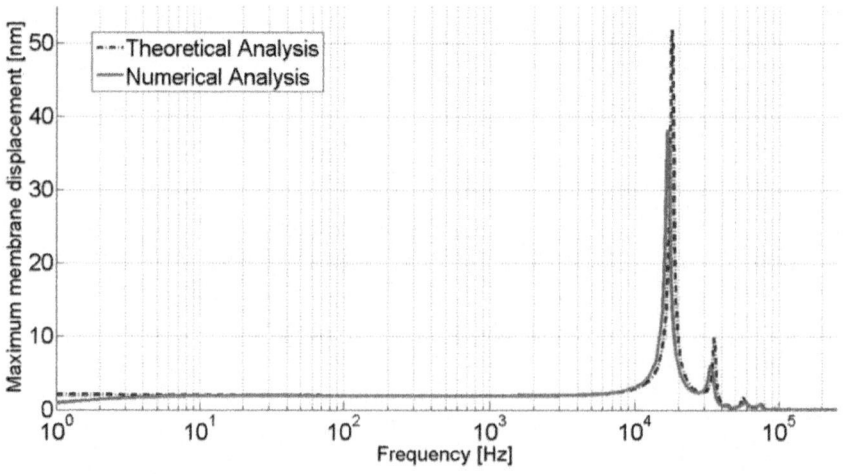

Figure 5. Comparison of the membrane displacement of theoretical and numerical analysis of the acoustic transducer (for Rm = 5 mm, R_b = 0.75 mm, h_c = 3000 μm, h_g = 30 μm, T_m = 500 N/m and t_m = 8 μm).

Selection of Parameters and Responses

Based on the theoretical analysis and the device structure, several parameters, such as membrane radius (R_m), bottom electrode radius (R_b), cavity height (h_c), air gap (h_g), membrane tension (T_m), membrane thickness (tm), materials properties (e.g., Young's modulus, density of the materials and Poisson ratio), and the venting hole geometry, are involved with device performance. During this study, geometry of the venting hole was kept unchanged and the polyethylene terephthalate (PET) thin film was used as a membrane material, whose properties are listed in Table 2. Thus venting-hole geometry and material properties were omitted from the further analysis.

The static capacitance of the system is generally determined by the effective surface area of the electrodes and air gap, whereas the quality factor depends on the damping loss mechanism that is related to the device geometry. On the other hand, the sensitivity of the acoustic sensor, such as capacitance variation, is basically driven by the membrane displacement. Therefore, the first step of this work is identification of major input parameters that strongly influence the membrane displacement and quality factor. In this regard, the classic one-variable-at-a-time method is used, where the effect of individual parameter on the membrane displacement at first resonance frequency was studied for a fixed set of other parameters at some nominal value using numerical simulation. The process is repeated for each of the parameters involved in the study until all the parameters have been studied.

It has been observed that the increase of the membrane displacement and quality factor, and shift of the resonance frequency are observed for the increasing membrane radius, as shown in Figure 6A. On the other hand, the increase of the bottom electrode radius leads to reduction of membrane displacement and quality factor (Figure 6B). Figure 6C illustrates the effect of cavity height on the membrane displacement. Large cavity height helps to reduce the air damping in the cavity and thus helps to increase the membrane displacement and quality factor. Similarly, increase in air gap provides higher membrane displacement and quality factor (Figure 6D); however, the increase in air gap leads to the lower static capacitance. In addition, as shown in Figure 6E, higher membrane tension reduces the membrane displacement, and also shifts the resonance at a higher frequency. Thickness of the membrane also affects the membrane displacement and the resonance frequency (Figure 6F).

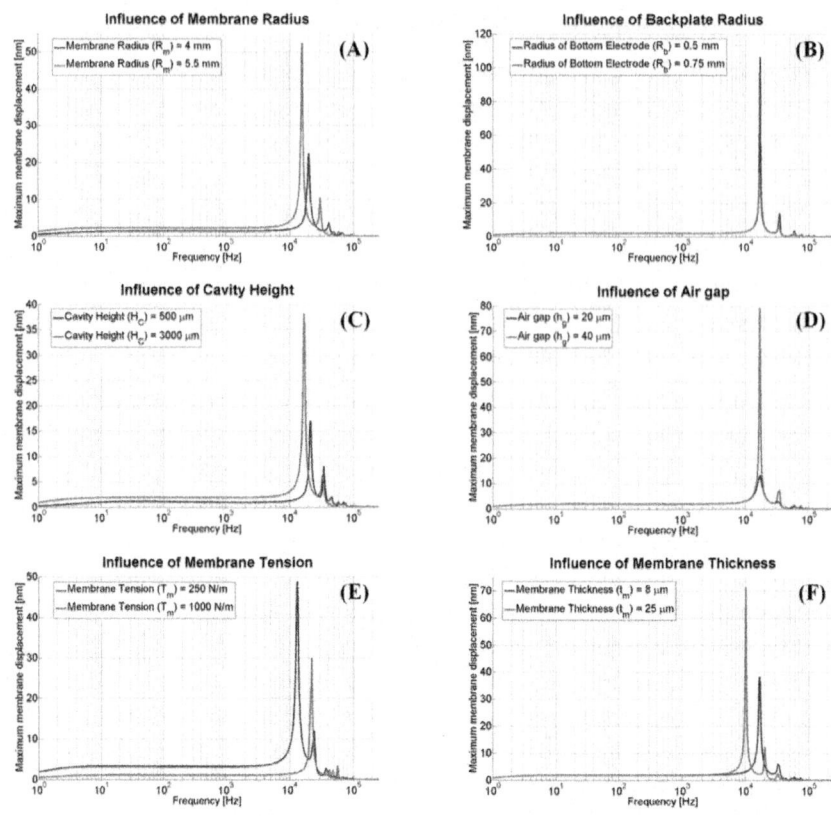

Figure 6: Effect of individual input parameter on the membrane displacement (other parameters kept at constant value: R_m = 5 mm, R_b = 0.75 mm, h_c = 3000 µm, h_g = 30 µm, T_m = 500 N/m and t_m = 8 µm).

Based on the initial tests, it has been observed that all six input parameters, namely R_m, R_b, h_c, h_g, T_m and t_m, have some influence on the membrane displacement and Q-factor, and therefore on the output responses. Moreover, to achieve better selectivity, the sensitivity at other natural frequencies than the first resonance frequency has to be reduced. Therefore, in order to study and eventually to optimize the capacitive acoustic resonator to fulfill the requirements, several output responses, specifically static capacitance (C_0), membrane displacement at first resonance ($|\xi_{Se}|_{fr1}$), quality factor (Q_f), capacitance variation (Δ_C), and membrane displacement at second resonance ($|\xi_{Se}|_{fr2}$) were studied for each experiment.

However, the one-variable-at-a-time approach cannot predict the interaction between the factors. In addition, this approach is not applicable for multiple response problems, and does not permit the construction of a model for the system [25]. Therefore, the study of the influence of all the input parameters and their interactions, and the optimization of the system requires methodical experimental strategies based on DOE. A good experimental strategy will provide the necessary information to estimate effects of factors and to develop empirical models for each system outputs and to optimize the multiple responses simultaneously to fulfill the objectives.

EXPERIMENTAL DESIGN

DOE provides a systematic way to study the effects of the input variables of a system or process, also known as factors, on outputs or responses. It is an effective tool for maximizing the amount of information gained from a study while minimizing the number of tests to be performed. In practice, DOE is applicable to both physical processes and numerical simulation models [26,27]; however, unlike physical measurement, numerical experimentation is not subject to noise or uncertainty [28,29].

Compared to other experimental strategies, namely one-variable-at-a-time and sequential simplex, RSM is intended to predict the response with a good quality all over the experimental domain. RSM approach has four basic steps: the data collection according to an experimental design, an empirical model (e.g., polynomial), calculation by least squares regression for each of the responses, generation of response surface contour plots or maps that are examined to the region of the desired response, and finally, the experimental verification of the predicted optimum [25]. DOE coupled with RSM can achieve rapid process development for minimal cost.

The selection of appropriate experiments is very important to build a reliable response surface model and therefore on its precise prediction [30]. According to the postulated model, there are different optimal design of experiments with a guarantee of good prediction in the domain of interest. One of the best known for a second-order model is the class of central composite design (CCD), consisting of a two-level complete or fractional factorial design, an "axial" design, and center points [30].

A series of FEM analyses of an acoustic sensor based on DOE have been performed to investigate the possibility of determining the optimal set of parameters to fabricate a sensor with optimum sensitivity and selectivity. The optimization is carried out to maximize the membrane displacement at first resonance frequency of the system, while minimizing the membrane displacement at other frequencies. The feasible domain is defined by the six factors, namely R_m, R_b, h_c, h_g, T_m, and t_m. The ranges of the six factors used in the numerical experiments are presented in Table 3; these values were selected based on the process capabilities of our equipment to fabricate devices.

Table 3: List of experimental variables (factors) and their ranges

Factors	Code	Range
Membrane Radius (R_m)	x_1	4–10 mm
Bottom Electrode Radius (R_b)	x_2	0.25–3 mm
Cavity Height (h_c)	x_3	1000–4000 μm
Air gap (h_g)	x_4	3–80 μm
Membrane Tension (T_m)	x_5	100–3000 N/m
Film Thickness (t_m)	x_6	8–25 μm

The variation domains of the six factors define a hypercube in six dimensions, and a second-order model was postulated to represent the evolution of the responses in this domain and optimize the acoustic resonant sensor. To estimate the coefficients of the model, a CCD was built with some additional points corresponding to a space-filling design to cover the entire domain. A total of 62 experiments were employed and listed in Table 4. These experiments were performed using numerical simulation, and for each experimental run, C0, $|\xi_{Se}|_{fr1}$, Qf, ΔC and $|\xi_{Se}|_{fr2}$ were collected for further analysis and empirical model building. During this study, "nemrodW" statistical software [31] is used to develop experimental strategies and search for optimal settings.

Table 4: DOE Table for acoustic sensor study

N°Exp	R_m	R_b	h_c	h_g	T_m	t_m	N°Exp	R_m	R_b	h_c	h_g	T_m	t_m
	mm	mm	μm	μm	N/m	μm		mm	mm	μm	μm	N/m	μm
1	4	0.25	1000	3	3000	8	32	10	3	4000	80	100	25
2	10	0.25	1000	3	100	8	33	10	0.25	1000	3	3000	25
3	4	3	1000	3	100	8	34	4	3	1000	3	3000	25
4	10	3	1000	3	3000	8	35	4	0.25	4000	3	3000	25
5	4	0.25	4000	3	100	8	36	10	3	4000	3	3000	25
6	10	0.25	4000	3	3000	8	37	4	0.25	1000	80	3000	25
7	4	3	4000	3	3000	8	38	10	3	1000	80	3000	25
8	10	3	4000	3	100	8	39	10	0.25	4000	80	3000	25
9	4	0.25	1000	80	100	8	40	4	3	4000	80	3000	25
10	10	0.25	1000	80	3000	8	41	10	1.625	2500	41.5	1550	16.5
11	4	3	1000	80	3000	8	42	7	0.25	2500	41.5	1550	16.5
12	10	3	1000	80	100	8	43	7	1.625	1000	41.5	1550	16.5
13	4	0.25	4000	80	3000	8	44	7	1.625	2500	3	1550	16.5
14	10	0.25	4000	80	100	8	45	7	1.625	2500	41.5	100	16.5
15	4	3	4000	80	100	8	46	7	1.625	2500	41.5	3000	16.5

16	10	3	4000	80	3000	8	47	5.9	1.322	2266	36.9	1407	15.8
17	4	1.625	2500	41.5	1550	8	48	8.1	1.322	2266	36.9	1407	15.8
18	10	1.625	2500	41.5	1550	8	49	7	2.231	2266	36.9	1407	15.8
19	7	0.25	2500	41.5	1550	8	50	7	1.625	3202	36.9	1407	15.8
20	7	3	2500	41.5	1550	8	51	7	1.625	2500	60.1	1407	15.8
21	7	1.625	1000	41.5	1550	8	52	7	1.625	2500	41.5	2265	15.8
22	7	1.625	4000	41.5	1550	8	53	7	1.625	2500	41.5	1550	20.8
23	7	1.625	2500	3	1550	8	54	4	0.5	1000	80	100	8
24	7	1.625	2500	80	1550	8	55	4	0.5	4000	80	3000	8
25	4	0.25	1000	3	100	25	56	10	0.5	4000	80	100	8
26	10	3	1000	3	100	25	57	7	0.5	2500	41.5	1550	8
27	10	0.25	4000	3	100	25	58	7	0.5	2500	80	1550	8
28	4	3	4000	3	100	25	59	4	0.5	4000	80	100	25
29	10	0.25	1000	80	100	25	60	4	0.5	1000	80	3000	25
30	4	3	1000	80	100	25	61	10	0.5	4000	80	3000	25
31	4	0.25	4000	80	100	25	62	7	0.5	2500	60.1	1407	15.8

RESULT AND DISCUSSION

Empirical Model Building and Analysis

To perform data analysis, the experimental data are first transformed into logarithmic scale to get symmetric distribution. The coefficients of the models are then estimated using common regression analysis techniques [27,32] to solve $X_{62\times28}\ \beta_{28\times1} = y_{62\times1}$, where X indicates the matrix of factors and factor interactions, vector y is the experimental results for one response in logarithmic scale, and vector β is the unknown coefficients. Generally, β is estimated by resolving the linear system of equations, and can be expressed as $\beta = (X^T X)^{-1} X^T y$, where "T" and "−1" represent the transpose and inverse matrix, respectively. Once the coefficients are computed, the equation of the empirical model for each response is entirely defined. For simplicity only the most significant terms of the empirical models are mentioned in the Equations (20) to (24) below, although each second order polynomial response equation consists of 28 terms.

$$Y_{CO} = 0.29743 + 0.09489x_1 + 0.61423x_2 - 0.71485x_4 - 0.04947x_1^2 - 0.09846x_2^2$$
$$+ 0.43464x_4^2 - 0.08399x_1x_2$$

$$(20)$$

$$Y_{|(\xi_{Se})|fr1} = 1.28557 + 0.22517x_1 - 0.87877x_2 + 0.25368x_3 + 0.95558x_4$$
$$- 0.24669x_5 + 0.11454x_6 + 0.51751x_2^2 - 0.46702x_4^2 - 0.29140x_5^2$$
$$+ 0.14770x_1x_2 + 0.18454x_1x_3 - 0.06673x_2x_3 - 0.19040x_2x_4$$
$$+ 0.09034x_1x_5 + 0.11223x_2x_5 - 0.06130x_3x_5 + 0.06152x_4x_6$$
$$+ 0.13720x_5x_6 \tag{21}$$

$$Y_{Qf} = 1.16077 - 0.45781x_2 + 0.09958x_3 + 0.12209x_4 + 0.09090x_6 + 0.75406x_2^2$$
$$- 0.31853x_3^2 + 0.24556x_1x_2 + 0.30512x_1x_3 - 0.19310x_2x_3 - 0.61056x_2x_4$$
$$- 0.16106x_3x_4 - 0.19291x_1x_5 + 0.15244x_2x_5 + 0.19565x_3x_5 + 0.36619x_4x_5$$
$$+ 0.12421x_2x_6 + 0.17754x_4x_6 \tag{22}$$

$$Y_{\Delta C} = -0.11475 + 0.24282x_1 + 0.16241x_2 + 0.26101x_3 - 0.44159x_4$$
$$- 0.24636x_5 + 0.11822x_6 + 0.64695x_4^2 + 0.15244x_1x_2 + 0.19087x_1x_3$$
$$- 0.20920x_2x_4 + 0.09150x_1x_5 + 0.11199x_2x_5 + 0.14269x_5x_6 \tag{23}$$

$$Y_{|(\xi_{Se})|fr2} = 0.43405 + 0.24954x_1 - 0.61280x_2 + 0.44395x_4 - 0.40514x_5$$
$$- 0.76316x_4^2 + 0.45812x_6^2 - 0.11145x_2x_3 + 0.21407x_2x_4$$
$$+ 0.14111x_2x_5 - 0.15529x_3x_5 - 0.24556x_4x_5 - 0.24635x_5x_6 \tag{24}$$

where Y_{C0}, $Y_{|\xi_{se}|}fr1$, Y_{Qf}, $Y_{\Delta C}$ and $Y_{|\xi_{se}|}fr2$ represent the empirical models of the responses for C0, $|\xi_{Se}|_{fr1}$, Qf, ΔC and $|\xi_{Se}|_{fr2}$, respectively, and x1, x2, x3, x4, x5, and x6 are the coded values of Rm, Rb, hc, h_g, Tm and tm, respectively.

Table 5: Analysis of variance (ANOVA) table of estimated models

Model	Source of Variation	Sum of Squares	Degrees of Freedom	Mean Square	Ratio	Sig.
$Y_{C0} = Log(C_0)$	Regression	42.2128	27	1.5634	6205.1558	<0.01
	Residuals	0.0076	30	0.0003		
	Total	42.2204	57			
	R-Squared (R^2)	1				
	Adj. R-Squared (R_a^2)	1				
$Y_{|\xi_{Se}|fr1} = Log(|\xi_{Se}|_{fr1})$	Regression	99.2762	27	3.6769	112.9190	<0.01
	Residuals	1.0094	31	0.0326		
	Total	100.2856	58			
	R-Squared (R^2)	0.990				
	Adj. R-Squared (R_a^2)	0.981				
$Y_Q = Log(Q)$	Regression	37.0305	27	1.3715	20.3269	<0.01
	Residuals	2.0242	30	0.0675		
	Total	39.0546	57			
	R-Squared (R^2)	0.948				
	Adj. R-Squared (R_a^2)	0.902				
$Y_{\Delta C} = Log(\Delta C)$	Regression	24.8283	27	0.9196	14.0307	<0.01
	Residuals	2.0317	31	0.0655		
	Total	26.8600	58			
	R-Squared (R^2)	0.924				
	Adj. R-Squared (R_a^2)	0.858				
$Y_{|\xi_{Se}|fr2} = Log(|\xi_{Se}|_{fr2})$	Regression	43.6303	27	1.6159	18.1856	<0.01
	Residuals	2.7546	31	0.0889		
	Total	46.3849	58			
	R-Squared (R^2)	0.941				
	Adj. R-Squared (R_a^2)	0.889				

To evaluate the significance of empirical models, analysis of variance (ANOVA) [33] is employed. The evaluated ANOVA of the model for logarithm of C0, logarithm of $|\xi_{Se}|_{fr1}$, logarithm of Qf, logarithm of ΔC, and logarithm of $|\xi_{Se}|_{fr2}$ are summarized in Table 5, respectively. The column "Sig." represents the p-values of the null hypothesis, which indicates the significance of the relation between factors and response, i.e., the model significance. With the p-values being less than 0.01 in all five models, it can be concluded that the five response surface models are statistically significant with 99% confidence level. Furthermore, the goodness of the fit of the regression model is measured by R-Squared (R2) and adjusted R-squared (R2a) values, which indicate the amount of variability in the response explained by the factors and range from 0 to 1. Therefore, the larger value is desirable. For all the models, R2-values are closer to 1, thereby indicating that the regression line perfectly fits the data. On the other hand, R2a-value provides the predictive accuracy. From the table, it has been observed that the value of R2 and R2a are very close to each other, suggesting that the models for all responses are adequately reproducing the experimental data. This approach ensures the inclusion of only those variables that have a significant effect in the statistical model.

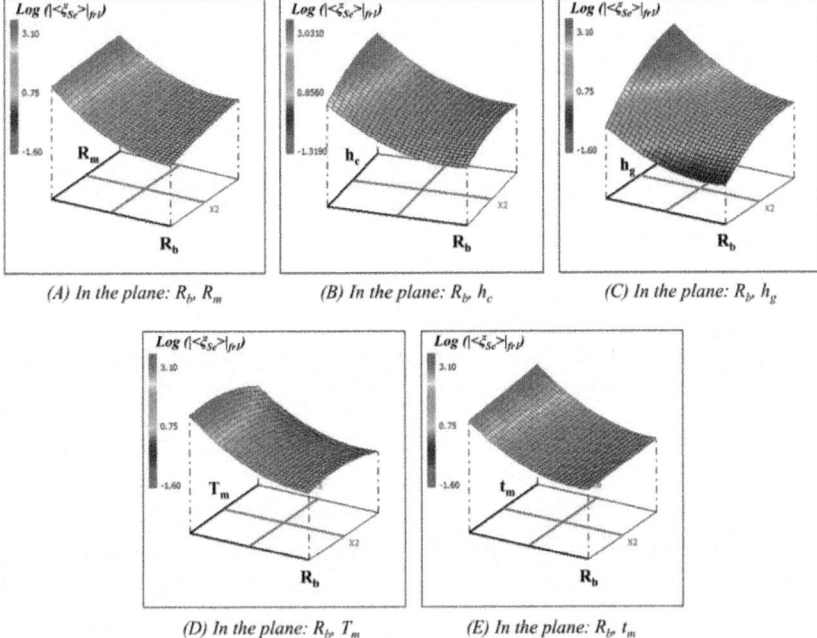

(A) In the plane: R_b, R_m   *(B) In the plane: R_b, h_c*   *(C) In the plane: R_b, h_g*

(D) In the plane: R_b, T_m   *(E) In the plane: R_b, t_m*

Figure 7: Response surface plots of logarithm of maximum membrane displacement at first resonance frequency (Log ($|\xi_{Se}|_{fr1}$)) in different planes with respect to other factors kept constants at the central values.

To further check the model behavior, response surface plots can provide a quick view to observe the maximum membrane displacement at first resonance frequency and the Q-factor for different values of factors and help to identify the type of interactions between these factors. Only two factors can be displayed on a plot while other factors are kept at constant levels at a central value. For example, the 3D graphical representations of the response surface of maximum membrane displacement at first resonance frequency and quality factor are illustrated in Figure 7 and Figure 8.

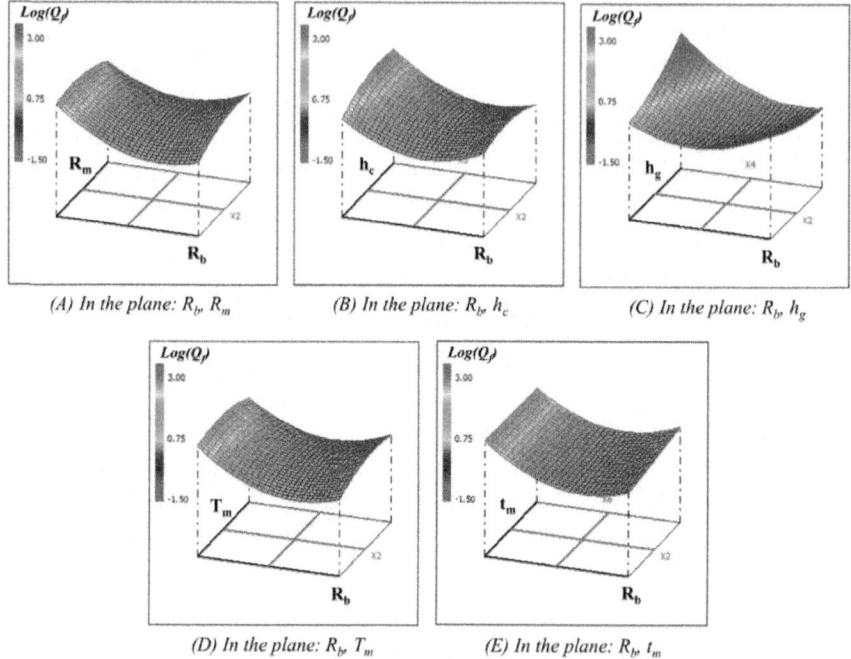

(A) In the plane: R_b, R_m (B) In the plane: R_b, h_c (C) In the plane: R_b, h_g

(D) In the plane: R_b, T_m (E) In the plane: R_b, t_m

Figure 8: Response surface plots of logarithm of quality factor ($Log(Q_f)$) in different planes with respect to other factors kept constant at the central values.

Figure 7A–E illustrate the interaction among different factors on the maximum membrane displacement at first resonance frequency for a fixed set of other factors at central values of their respective variation domain. It has been observed that all the six parameters, and some of the quadratic, as well as interaction between those parameters, have the strongest influence on the membrane displacement as presented by Equation (21). On the other hand, Figure 8A–E show the interaction of different factors on the quality factor of the membrane displacement at first resonance while other factors fixed at central levels. Observation reveals that the Q-factor has been influenced by linear terms R_b, h_c, h_g, t_m, quadratic terms R^2_b, h^2_c, and interaction terms $R_m R_b$,

$R_m h_c$, $R_m T_m$, $R_b h_c$, $R_b h_g$, $R_b T_m$, $R_b t_m$, $h_c h_g$, $h_c T_m$, $h_g T_m$, $h_g t_m$ etc. The response surface plots also show the local maxima and minima of the responses in terms of different factors within their investigated ranges. As an example, membrane displacement of an acoustic resonator can be maximized by increasing the value of R_m, h_c, h_g and t_m, and by reducing the value of R_b and T_m as illustrated by red color zone in Figure 7.

Optimization Process

As observed, the effects of factors are not only additive but also interactive. The presence of interaction effects makes it imperative that all the factors be optimized simultaneously to determine the best compromise and multi-criteria optimization is necessary. Desirability function approach is employed to achieve simultaneous optimization in our multi-response problems. In this approach, an objective function, also known as desirability function, is used to transform the existing values of the considered response in to a scale-free value called desirability. The desirability lies between 0 and 1 and it represents the closeness of a response to its ideal value.

Multi-response optimization problem generally involves several processing steps after the models being fitted with the experimental data. Initially, the desirability index (di) was defined for each response, based on the part of desirability function as presented in Equations (25) to (27), for the cases of bilateral desirability function, maximization and minimization [11,15].

$$d_i(\hat{Y}_i) = \begin{cases} 0 & \text{if } \hat{Y}_i < a \\ \left(\dfrac{\hat{Y}_i - a}{a_i - a}\right)^{w_{i1}} & \text{if } a < \hat{Y}_i < a_1 \\ 1 & \text{if } a_1 < \hat{Y}_i < b_1 \\ \left(\dfrac{b - \hat{Y}_i}{b - b_1}\right)^{w_{i2}} & \text{if } b_1 < \hat{Y}_i < b \\ 0 & \text{if } \hat{Y}_i > b \end{cases} \tag{25}$$

$$d_i(\hat{Y}_i) = \begin{cases} 0 & \text{if } \hat{Y}_i < a \\ \left(\dfrac{\hat{Y}_i - a}{b - a}\right)^{w_i} & \text{if } a < \hat{Y}_i < b \\ 1 & \text{if } \hat{Y}_i > b \end{cases} \tag{26}$$

$$d_i(\hat{Y}_i) = \begin{cases} 1 & \text{if } \hat{Y}_i < a \\ \left(\dfrac{b - \hat{Y}_i}{b - a}\right)^{w_i} & \text{if } a < \hat{Y}_i < b \\ 0 & \text{if } \hat{Y}_i > b \end{cases} \tag{27}$$

where "a" represents the lower tolerance limit, "b" represents the upper tolerance limit, and "a1 and b1" represent the target interval. The w_i, w_{i1} and w_{i2} in Equations (25) to (27) represent the considered weights. Shape of desirability functions are respectively illustrated in Figure 9.

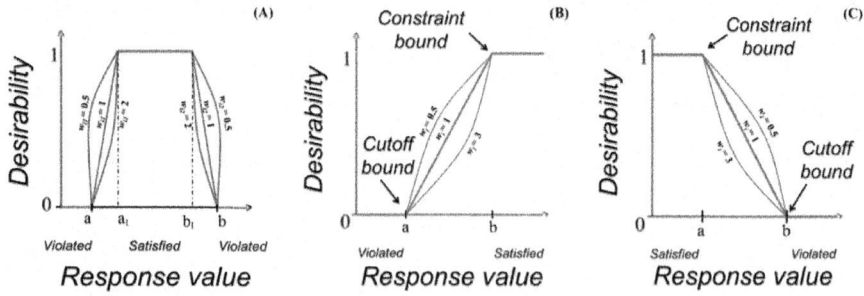

Figure 9: Schematic diagram of different desirability functions: **(A)** bilateral desirability function, **(B)** maximization and **(C)** minimization.

Global desirability is then calculated by accumulating the "n" individual desirability values corresponding to the "n" studied responses, as follows,

$$ D = \left(\prod_{i=1}^{n} d_i^{w_i} \right)^{1/\sum w_i} $$

(28)

Here, D is the global desirability, d_i' represents the respective individual desirability, and $\sum w_i$ represents the total weight.

Thereafter, the optimum combination of levels of parameters is determined based on the highest global desirability value. Finally, the response of the sensor based on the optimum level of parameters is predicted and validated.

In this study, the optimization is performed to obtain an acoustic resonant sensor, whereby the Q-factor and capacitance variation are maximized, while the value of static capacitance is held within the fixed value range and the membrane displacement at second resonance frequency is minimized to achieve better selectivity. Table 6 represents the list of optimization criteria and the desirability functions for responses that have been used during the optimization process. Figure 10A–D illustrate the desirability functions.

Table 6: Optimization criteria and desirability functions for the optimization of an acoustic resonant sensor

Response (unit)	Partial Desirability Code	Functions	Weight (w_i)	a	b	Predicted Response	Partial Desirability
C_0 (pF)	d_1	Bilateral	1	0.5	3.2	0.5	100%
Q_f	d_2	Maximization	1	25	1450	210	52.4%
ΔC (fF)	d_3	Maximization	1	1	36	1.72	15.1%
$|<\zeta_{Se}^z>|_{fr2}$ (nm)	d_4	Minimization	1	0.03	3	1.12	21.3%

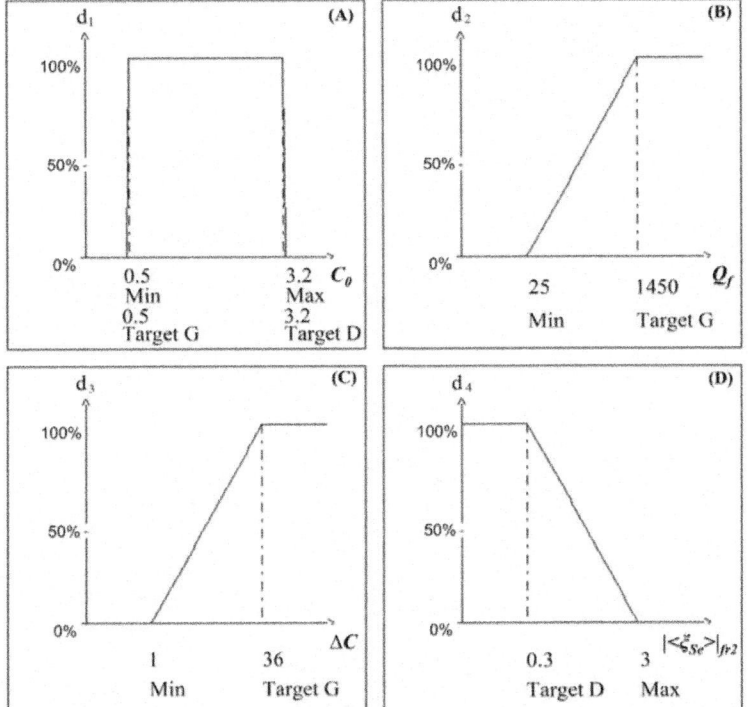

Figure 10: Desirability functions for multi-criteria optimization of acoustic resonant sensor (for (**A**) $0.5 \leq C0 \leq 3.2$ pF; (**B**) $Qf \geq 25$; (**C**) $\Delta C \geq 1$ fF; (**D**) $|\xi_{Se}|_{fr2} \leq 3$ nm).

The solution found based on multi-criteria optimization is presented by response surface of global desirability with respect to different planes in Figure 11A–E, where white region represents the acceptable zone that satisfies all the criteria. Finally, the global desirability is evaluated based on which optimum level of parameters is decided to satisfy the desirability. The estimated set of optimized parameters based on multi-criteria optimization is listed in Table 7, which provides the global desirability of 36%.

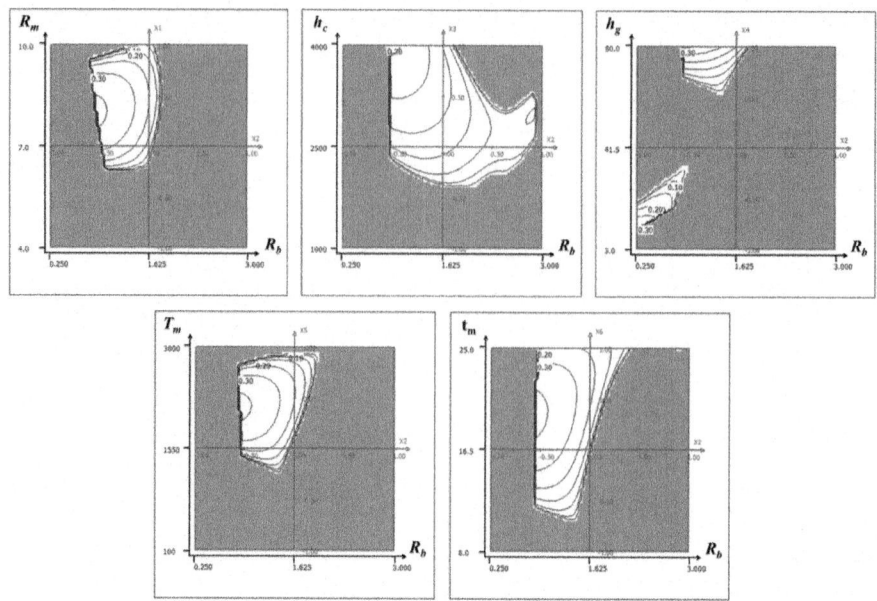

Figure 11: Optimum zone for acoustic resonant sensor with desired responses.

Table 7: Set of optimized parameters based on multi-criteria desirability functions optimization.

Factor	Value
Membrane radius (R_m)	8.1 mm
Backplate radius (R_b)	0.871 mm
Cavity height (h_c)	3987 µm
Air gap (h_g)	80.0 µm
Membrane tension (T_m)	2158 N/m
Membrane thickness (t_m)	19.8 µm

Verification

Once the optimum set of parameters is determined, the numerical analysis has been performed to verify the responses of the acoustic sensor with optimum parameters. Figure 12 shows the maximum membrane displacement of the acoustic resonant sensor with respect to frequencies. It has been observed according to numerical analysis that the acoustic sensor with a set of optimum parameters provides good sensitivity and selectivity, with static capacitance (C0) of 0.50 pF, capacitance variation (ΔC) of 2.6 f$_F$, and Q-factor (Q$_f$) of 522, alone with capacitance ratio ($\Delta C/C0$) of 0.52% at first resonance frequency of the acoustic sensor for an incident acoustic pressure level of (or equal to) 80 dB_{SPL}.

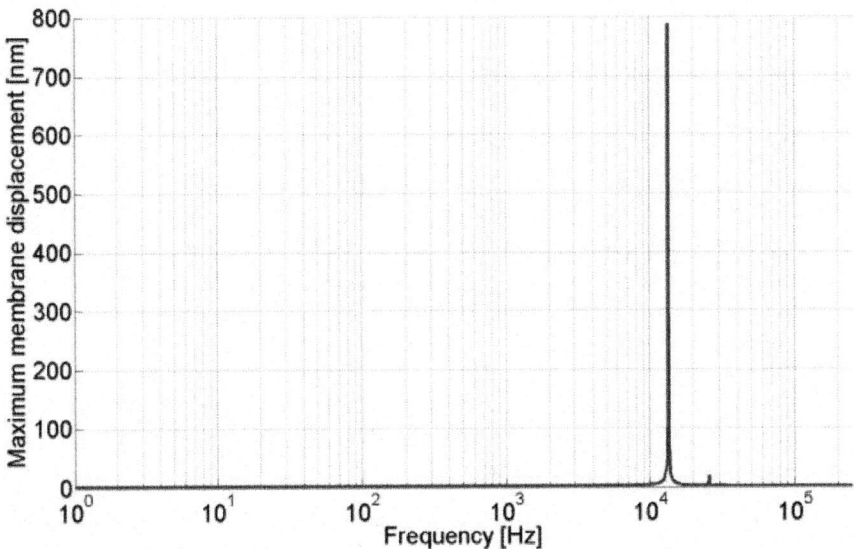

Figure 12: Maximum membrane displacement of the acoustic resonant sensor with set of optimum parameters.

CONCLUSIONS

Numerical simulation and the DOE approach can be used to investigate the virtual prototyping of an acoustic sensor to understand the linear, quadratic, and interaction effects of different parameters on the outputs of the sensor. DOE helps to reduce the computation efforts in the acoustic resonant sensor optimization process since the empirical model is far less complex than the numerical simulation. RSM helps to develop empirical model for each response. It has been observed that the maximum membrane displacement at first resonance frequency and quality factor are influenced by several linear, quadratic, and interaction terms. Based on the empirical model, the region of the optimum set of parameters for an acoustic resonant sensor was obtained using multi-criteria optimization. During this work, global desirability of 36% was achieved. Cross-verification using numerical simulation shows that the capacitance of 0.50 pF, capacitance variation of 2.6 fF, and quality factor of 522 can be achieved. Hence, the optimum set of parameters satisfies the targeted output response of the acoustic resonator.

ACKNOWLEDGMENTS

The authors are thankful to Petr Honzík and Stéphane Durand for the productive discussions related to acoustic. The authors are also grateful to SPINNAKER project for the funding.

AUTHOR CONTRIBUTIONS

Rubaiyet Haque performed the experiments, analyzes the data and wrote the paper; Michelle Sergent proposed the experimental design and introduce the DOE software the experiments; Christophe Loussert provides the industrial requirements and guidance; Xavier Boddaert and Patrick Benaben supervised the work.

REFERENCES

1. Martin, D.T. Design, Fabrication, and Characterization of a MEMS Dual-backplate Capacitive Microphone. Ph.D. Thesis, University of Florida, Gainesville, FL, USA, 2007.

2. Chatzopoulos, D. Modeling the Performance of MEMS Based Directional Microphones. Master's Thesis, Naval Postgraduate School, Monterey, CA, USA, 2008.

3. Kaushik, B.; Nance, D.; Ahuja, K.K. A Review of the Role of Acoustic Sensors in the Modern Battlefield. In Proceedings of the 11th AIAA/ CEAS Aeroacoustics Conference, Monterey, CA, USA, 23–25 May 2005.

4. Scheeper, P.R.; van der Donk, A.G.H.; Olthuis, W.; Bergveld, P. A Review of Silicon Microphones. Sens. Actuators A Phys. **1994**, 44, 1–11.

5. Hsu, P.-C.; Mastrangelo, C.H.; Wise, K.D. A High Sensitivity Polysilicon Diaphragm Condenser Microphone. In Proceedings of the 11th Annual International Workshop on Micro Electro Mechanical Systems Proceeding, Heidelberg, Germany, 25–29 January 1998; pp. 580–585.

6. Hohm, D.; Hess, G. A Subminiature Condenser Microphone with Silicon Nitride Membrane and Silicon Back Plate. J. Acoust. Soc. Am. **1989**, 85, 476–480.

7. Shu, Z.-Z.; Ke, M.-L.; Chen, G.-W.; Horng, R.-H.; Chang, C.-C.; Tsai, J.-Y.; Lai, C.-C.; Chen, J.-L. Design and Fabrication of Condenser Microphone Using Wafer Transfer and Micro-Electroplating Technique. In Proceedings of the Design, Test, Integration and Packaging of MEMS/ MOEMS, Nice, France, 9–11 April 2008; pp. 386–390.

8. Zou, Q.; Li, Z.; Liu, L. Design and Fabrication of Silicon Condenser Microphone Using Corrugated Diaphragm Technique. J. Microelectromech. Syst. **1996**, 5, 197–204.

9. Honzík, P.; Podkovskiy, A.; Durand, S.; Joly, N.; Bruneau, M. Analytical and Numerical Modeling of an Axisymmetrical Electrostatic Transducer with Interior Geometrical Discontinuity. J. Acoust. Soc. Am. **2013**, 134, 3573–3579.

10. Podkovskiy, A.; Honzík, P.; Durand, S.; Joly, N.; Bruneau, M. Miniaturized Electrostatic Receiver with Small-Sized Backing Electrode. In Proceedings of Meetings on Acoustics, Montreal, QC, Canada, 2–7 June 2013; Volume 19.

11. Lewis, G.A.; Mathieu, D.; Phan-Tan-Luu, R. Pharmaceutical Experimental Design; Dekker: New York, NY, USA, 1999.

12. Vogel, F.; Landes, H.; Lerch, R.; Kaltenbacher, M.; Peipp, R. Numerical Simulation and Optimization of Capacitive Transducers. In Proceedings of EuroSime, Aix-en-Provence, France, 30 March–2 April 2003; pp. 399–405.

13. Gill, P.E.; Wong, E. Sequential Quadratic Programming Methods. In Mixed Integer Nonlinear Programing; Lee, J., Leyffer, S., Eds.; Springer: New York, NY, USA, 2012; Volume 154, pp. 147–224.

14. Derringer, G.; Suich, R. Simultaneous Optimization of Several Response Variables. J. Qual. Technol. **1980**, 12, 214–219.

15. Sarabia, L.A.; Ortiz, M.C. Response Surface Methodology. In Comprehensive Chemometrics; Steven, B., Tauler, R., Walczak, B., Eds.; Elsevier: Oxford, UK, 2009; Volume 1, pp. 345–390.

16. Lavergne, T.; Durand, S.; Bruneau, M.; Joly, N. Dynamic Behaviour of the Circular Membrane of an Electrostatic Microphone: Effect of Holes in the Backing Electrode. J. Acoust. Soc. Am. **2010**, 128, 3459–3477.

17. Bruneau, M.; Bruneau, A.-M.; Škvor, Z.; Lotton, P. An Equivalent Network Modelling the Strong Coupling Between a Vibrating Membrane and a Fluid Film. Acta Acust. **1994**, 2, 223–232.

18. Bower, A.F. Applied Mechanics of Solids; CRC Press: New York, NY, USA, 2010.

19. Morse, P.M.; Uno Ingard, K. Theoretical Acoustics; Princeton University Press: Princeton, NJ, USA, 1986.

20. Merhaut, J. A Contribution to the Theory of Electroacoustic Transducers Based on Electrostatic Principle. Acustica1967, 19, 283–292.

21. Baker, W.P.; Kriegsmann, G.A.; Reiss, E.L. Acoustic Scattering by Baffled Cavity-Backed Membranes. J. Acoust. Soc. Am. **1988**, 83, 423–

432.

22. Prak, A.; Blom, F.R.; Elwenspoek, M.; Lammering, T.S.J. Q-Factor and Frequency Shift of Resonating Silicon Diaphragms in Air. Sens. Actuators A Phys. **1991**, 25–27, 691–698.

23. Ren, S.; Yuan, W.; Qiao, D.; Deng, J.; Sun, X. A Micromachined Pressure Sensor with Integrated Resonator Operating at Atmospheric Pressure. Sensors **2013**, 13, 17006–17024.

24. Park, K.K.; Lee, H.J.; Crisman, P.; Kupnik, M.; Oralkan, O.; Khuri-Yakub, B.T. Optimum design of circular CMUT membranes for high quality factor in air. In Proceddings of the Ultrasonics Symposium, Beijing, China, 2–5 November 2008; pp. 504–507.

25. Rautela, G.S.; Snee, R.D.; Miller, W.K. Response-Surface Co-optimization of Reaction Conditions in Clinical Chemical Methods. Clin. Chem. **1979**, 25, 1954–1964.

26. Telford, J.K. A Brief Introduction to Design of Experiments. Johns Hopkins APL Tech. Dig. **2007**, 27, 224–232.

27. Bahloul, R.; Mkaddem, A.; Dal Santo, P.; Potiron, A. Sheet Metal Bending Optimisation using Response Surface Method, Numerical Simulation and Design of Experiments. Int. J. Mech. Sci. **2006**, 48, 991–1003.

28. Stehouwer, P. Design of Experiments for Numerical Parameter Studies of Electronic Systems: Optimizing the Cooling Strategy of an Ethernet Switch. Electronics Cooling Magazine, May 2005.

29. Gou, J.; Zhang, C.; Liang, Z.; Wang, B.; Simpson, J. Resin Transfer Molding Process Optimization Using Numerical Simulation and Design of Experiments Approach. Polym. Compos. **2003**, 24, 1–12.

30. Rosales, E.; Sanromán, M.A.; Pazos, M. Application of Central Compostie Face-Centered Design and Response Surface Methodology for the Optimization of Electro-Fenton Decolorization of Azure B Dye. Environ. Sci. Pollut. Res. **2012**, 19, 1738–1746.

31. LPRAI. nemrodW. Available online: http://www.lprai.com/index.php?page=Logiciel (accessed on 27 December 2014).

32. Montgomery, D.C.; Runger, G.C. Multiple Linear Regression. In Applied Statistics and Probability for Engineers, 4th ed.; John Wiley & Sons: New York, NY, USA, 2007; pp. 410–467.

33. Montgomery, D.C.; Runger, G.C. Design and Analysis for Single-Factor Experiments: The Analysis of Variance. In Applied Statistics and Probability for Engineers, 4th ed.; John Wiley & Sons: New York, NY, USA, 2007; pp. 468–504.

Chapter 6

EFFICIENT NUMERICAL METHODS FOR SOLVING DIFFERENTIAL ALGEBRAIC EQUATIONS

Ampon Dhamacharoen

Department of Mathematics, Burapha University, Chonburi, Thailand

ABSTRACT

This research aims to solve Differential Algebraic Equation (DAE) problems in their original form, wherein both the differential and algebraic equations remain. The Newton or Newton-Broyden technique along with some integrators such as the Runge-Kutta method is coupled together to solve the problems. Experiments show that the method developed in this paper is efficient, as it demonstrates that implementation of the method is not difficult, and such method is able to provide approximate solutions with ease within some desired accuracy standards.

INTRODUCTION

A differential algebraic Equation (DAE) is an equation involving an unknown function and its derivatives. A DAE in its most general form is given by the following:

$$F\left(t, x(t), y(t), y'(t)\right) = 0, \quad t \in [a, b] \tag{1}$$

where R is the set of real number, $F: R \times R^m \times R^n \times R^n \to R^{m+n}$, $x(t) \in R^m$, $y(t) \in R^n$ and $y'(t) \in R^n$. In this form, the relationship between the variables and derivatives may be implicit. In some systems, the equations may be written in the explicit form of derivatives, as follows:

$$y'(t) = f\left(t, x(t), y(t)\right) \tag{2a}$$

$$g\left(t, x(t), y(t)\right) = 0 \tag{2b}$$

where $f: R \times R^m \times R^n \rightarrow R^n$, This set of equations is called a semi-explicit DAE system.

Differential algebraic equations arise in the mathematical modeling of a wide variety of problems found in engineering and science, such as multi-body and flexible body mechanics, electrical circuit design, optimal control, incompressible fluids, molecular dynamics, chemical kinetics and chemical process control [1] -[5] .

DAEs can be transformed into ODE problems via differentiation. The number of differentiations needed in the transforming process is called the differentiation index. This number can describe some characteristics of the problem. In general, the higher the index of a DAE, the more difficulties one can expect in its numerical solution [1] - [3] [6] .

Although DAEs can be transformed into an explicit ODE so that it can be solved using the methods of ODE, there are still many numerical methods that can solve DAE directly. In the DAE solvers software, the numerical approaches for the solution of DAEs can be divided roughly into two classes: a) direct discretizations of the given system; and b) methods which involve a reformulation (e.g. index reduction), combined with a discretization. Direct discretizations are easier to use, but are limited in their utilitisation to essentially index-1, index-2 Hessenberg, and index -3 Hessenberg DAE systems, while a reformulation may be costly, and it may also require more input from the user and involve more user intervention [3] [6].

In this research, we will place emphasis on constructing an algorithm for solving a semi-explicit DAE. The approach employed is to firstly solve the system of algebraic equations, and subsequently solve the differential equations using the derived information. The Newton-Broyden method plays a key role in solving algebraic equations, since it performs almost as well as the Newton method, but requires less energy, and in addition, it also outperforms other methods of the same order of convergence [7] - [9] .

THE NEWTON-BROYDEN METHOD

The method, which was first proposed by Dhamacharoen in 2011 [7] , [8] , aims to solve the equation of the form:

$$F\left(u\left(x\right),x\right)=0 \tag{3}$$

The Newton scheme of this problem is:

$$x_{i+1} = x_i - \left[F_u\left(u\left(x_i\right),x_i\right)u'\left(x_i\right) + F_x\left(u\left(x_i\right),x_i\right)\right]^{-1} F\left(u\left(x_i\right),x_i\right), \quad i=0,1,2,\cdots. \tag{4}$$

Replacing $u'(x_i)$ by D_i, gives

$$x_{i+1} = x_i - \left[F_u\big(u(x_i), x_i\big) D_i + F_x\big(u(x_i), x_i\big) \right]^{-1} F\big(u(x_i), x_i\big), \quad i = 0,1,2,\cdots. \tag{5}$$

with updating:

$$D_{i+1} = D_i - \frac{1}{b_i^T b_i}\big(u(x_{i+1}) - u(x_i) - D_i b_i\big) b_i^T \tag{6}$$

where $b_i = x_{i+1} - x_i$. Equation (6) is called the Broyden rank-1 update, and (5) with the update (6) is called the Newton-Broyden method. This method retains the good part of the Newton method, and replaces the difficult part of Newton's with Broyden's. With good initial guesses for z_0 and D_0, the Newton-Broyden Scheme (5) will produce a sequence that converges to a solution of (3), with q-super linear order of convergence.

Although the order of convergence of the Newton-Broyden method is equal to that of the Broyden method, and less than that of the Newton method, in practice, the Newton-Broyden performs well in the sense that a good initial guess is easily found, and it reaches the solution in a reasonable number of ierations. As shown in [7] and [8] , for the same problem and using the same initial guess, the sequences from the Newton method and from the Newton-Broyden method reach the solution, while that from the Broyden does not. In addition, the Newton-Broyden method requires less amount of work in comparison with the Newton method.

Note that if function F is linear, then Broyden's update and Newton-Broyden's update coincide.

CONSTRUCTING THE METHOD

The initial value problems:

Assume that the DAE is expressed in the form (2a), (2b) (renumbering)

$$y'(t) = f\big(t, x(t), y(t)\big) \tag{7a}$$

$$g\big(t, x(t), y(t)\big) = 0 \tag{7b}$$

$$y(t_0) = y_0 \tag{7c}$$

in which the system of equations is to be solved for x(t) and y(t), where t Î [a, b]. In solving the system numerically, the conditions on the initial values of y must be imposed sufficiently for the system to have a unique solution. If (7b) can express x(t) in terms of t and y(t) explicitly, then the system becomes a pure ODE. Therefore, we consider the case when (7b) expresses x(t) implicitly.

Let the interval [a, b] be divided into n subintervals $\{a = t_0, t_1, \cdots, t_n = b\}$. In each interval $[t_{i-1}, t_i]$, we will solve (7a) numerically for $y(t_i)$. In an initial value problem, the value $y(t_0)$ is specified, and the value $x(t_0)$ can subsequently be solved from (7b). In order to use the Runge-Kutta method, in each interval $[t_{i-1}, t_i]$ the value x_1, x_2 and x_3, which are needed for computing F_2, F_3 and F_4 respectively, can each be solved from (7b) using the values of $y(t_{i-1}) + F_1/2$ in F_2, $y(t_{i-1}) + F_2/2$ in F_3 and $y(t_{i-1}) + F_3/2$ in F_4. Once the value $y(t_i)$ is computed, $x(t_i)$ can then be solved from (7b). Advance to the next interval, and repeat this procedure until we reach the last subinterval.

Suppose the term x(t) is missing from the Equation (7b). Therefore, the system becomes:

$$y'(t) = f(t, x(t), y(t))$$

$$\text{(8a)}$$

$$g(t, y(t)) = 0$$

$$\text{(8b)}$$

$$y(t_0) = y_0$$

$$\text{(8c)}$$

in which the system is called index-2 Hessenberg. In solving the differential Equation (8a), the variable x(t) are treated as unknowns. In each interval $[t_{i-1}, t_i]$, the value of $x(t_{i-1})$ is given as a guess, and (8a) is then solved numerically for $y(t_i)$. Check the condition (8b) $g(t_i, y(t_i)) = 0$. Update the value of $x(t_0)$ and iterate the process until the condition (8b) is met. Subsequently, advance to the next interval using the last value of $x(t_{i-1})$ as the guessing value for $x(t_i)$. Repeat this procedure until we complete the last subinterval.

In solving the differential equations, we may use the 4th order Runge-Kutta method, or the 4th order Taylor method, since their local error is $O(h^5)$ and the total error is $O(h^4)$, which is acceptable and also easy to implement. In solving the algebraic equations associated in the problem, the Newton method is used in (7b), and the Newton-Broyden method is used in (8b).

As per the procedure described above, the formulation will be as follows:

Partition the interval [a, b] into n subintervals. $h = \dfrac{b-a}{n}$, and

$t_0 = a, \ t_i = t_{i-1} + h, \ i = 1, 2, \cdots, n.$

Problem (7a), (7b):

Define the function F: $F\left(x\left(t_{i-1}\right)\right) = g\left(t_i, x\left(t_i\right), y\left(t_i\right)\right), i = 1, 2, \cdots, n \cdot$

The problem become (3)

$F\left(x\left(t_{i-1}\right)\right) = 0$

which can be solved using the Newton method.

P1: Newton Method.

Prescribe a small positive real number e.

Guess the value z.

A: Compute F(z).

Check if $\|F(z)\| < e$

If so, proceed to B. If not, carry out the next step.

Compute F'(z).

Solve the system F'(z)b = −F(z), for b.

Set z = z + b.

Go to A.

B: End.

The process for solving the DAE will be as follows:

Algorithm A:

Initial step:

1) Set the given initial condition $y(t_0) = y_0$.

2)   Solve for $x(t_0)$, using P1: Set $x(t_0) = z$.

Main step:

For $i = 1, 2, \cdots, n$; process the following steps:

Step 1. Solve the initial value problem (7a) 1 step to obtain $y(t_i)$, by proceeding using the following steps:

1)   Compute: $F_1 = hf\left(t_{i-1}, x\left(t_{i-1}\right), y\left(t_{i-1}\right)\right)$, $y_1 = y\left(t_{i-1}\right) + F_1/2$, Solve $g\left(t_{i-1} + h/2, x_1, y_1\right) = 0$ for x_1. (Using P1).

2)   Compute: $F_2 = hf\left(t_{i-1} + h/2, x_1, y_1\right)$ $y_2 = y\left(t_{i-1}\right) + F_2/2$, Solve $g\left(t_{i-1} + h/2, x_2, y_2\right) = 0$ for x_2. (Using P1).

3) Compute: $F_3 = hf\left(t_{i-1} + h/2, x_2, y_2\right)$ $y_3 = y(t_{i-1}) + F_3/2$, Solve $g\left(t_{i-1} + h, x_3, y_3\right) = 0$ for x_3. (Using P1).

4) Compute: $F_4 = hf\left(t_{i-1} + h, x_3, y_3\right)$

5) $y(t_i) = y(t_{i-1}) + \left(F_1 + 2\left(F_2 + F_3\right) + F_4\right)/6$

Step 2. Solve (7b) for $x(t_i)$, using P1: Set $x(t_i) = z$.

Proceed to the next interval.

Problem (8a), (8b):

Define the function F: $F\left(x(t_{i-1}), y(t_{i-1})\right) = g\left(t_i, y(t_i)\right), i = 1,2,\cdots,n$.

The problem become (3)

$$F\left(x(t_{i-1}), y(t_{i-1})\right) = 0$$

which can be solved using the Newton-Broyden method.

P2: One Step Runge-Kutta Method Given the value $t_{i-1}, x(t_{i-1}), y(t_{i-1})$ and b.

Compute: $F_1 = hf\left(t_{i-1}, x(t_{i-1}), y(t_{i-1})\right)$, $y_1 = y(t_{i-1}) + F_1/2$, $x_1 = x(t_{i-1}) + b/2$.

Compute: $F_2 = hf\left(t_{i-1} + h/2, x_1, y_1\right)$, $y_2 = y(t_{i-1}) + F_2/2$, $x_2 = x(t_{i-1}) + b/2$.

Compute: $F_3 = hf\left(t_{i-1} + h/2, x_2, y_2\right)$, $y_3 = y(t_{i-1}) + F_3$, $x_3 = x(t_{i-1}) + b$.

Compute: $F_4 = hf\left(t_{i-1} + h, x_3, y_3\right)$.

Then $y(t_i) = y(t_{i-1}) + \left(F_1 + 2\left(F_2 + F_3\right) + F_4\right)/6$.

The process for solving the DAE will be as follows:

Algorithm B:

Initial step:

1) Set the given initial condition $y(t_0) = y_0$.

2) Guess the initial condition $x(t_0) = z$, $x(t_0) = b$.

3) Guess the initial matrix D (may be = I), Guess the value b.

4) Solve the initial value problem (8a) 1 step, to obtain $y(t_1)$, using P2.

5) Compute the value $F\left(x(t_0), y(t_0)\right) = g\left(t_1, y(t_1)\right)$.

Main step:

For $i = 1,2,\cdots,n$; process the following steps:

Step 1. Check the condition $\|g(t_i, y(t_i))\| < \varepsilon$ where e is a prescribed small number.

If it passes, proceed to the next interval (next i).

If it fails, carry out step 2.

Step 2.

1) Set $u = y(t_i)$.

2) Compute the matrix $A = F_x\left(x(t_{i-1}), y(t_i)\right) + F_y\left(x(t_{i-1}), y(t_i)\right) D.$

3) Solve $Ab = -g\left(t_i, y(t_i)\right)$ for b.

4) Set $x(t_{i-1}) = x(t_{i-1}) + b.$

5) Solve the initial value problem (8a) 1 step (using P2) to obtain $y(t_i)$.

6) Compute the new value $g\left(t_i, y(t_i)\right).$

7) Update the matrix $D = D + \dfrac{1}{b^T b}\left(y(t_i) - u - Db\right)b^T.$

8) Go to step 1.

End.

EXPERIMENTS

Three examples are used to illustrate the method. The first problem is an index-1 Hessenberg DAE system, with nonlinear differential equations and a nonlinear algebraic equation, as follows: $x''(t) = -(3t+1)y(t) - x(t)(4z(t)+1), \ 0 \le t \le 1$

$y''(t) = 4\cos(z(t)) - y(t)(4z(t)+1)$ initial conditions,

$x(0) = 0, \ y(0) = 0, \qquad x'(0) = 1, \ y'(0) = 2$ algebraic equation

$4x(t)\cos(z(t)) + ty^2(t) = 4\left(z(t) - t^2\right)$

Using Algorithm A, this problem is solved and has a nice result with a small error as compared to the exact solution $z(t) = t(t+1) \quad x(t) = t\cos(z(t))$ $y(t) = \sin(z(t))$.

Some results are illustrated in Table 1. (xs, ys and zs are values from the exact solution).

The second problem is a classical example of the DAE problem which is "The pendulum problem", as expressed in the xy co-ordinate plane as follows:

$$\left. \begin{array}{l} x''(t) = -l(t)x(t), \quad t \geq 0 \\ y''(t) = -l(t)y(t) - g \\ x(t)^2 + y(t)^2 = L^2 \end{array} \right\}$$

(8)

where g = 9.8. This problem is an index 2 semi-explicit DAE problem. [3] , [10] . Proceeding to solve the problem numerically, we let L = 1, and impose the initial conditions $x(0) = 1$, $y(0) = 0$. Using the new variables $y_1 = x$, $y_2 = x'$, $y_3 = y$ and $y_4 = y'$, we have a system of differential algebraic equations as follows:

Table 1: Resulting values for x(t), y(t) and z(t) from Algorithm A, and the errors

i	t	x	y	z	x ? xs	y ? ys	z ? zs
0	0.000000	0.000000	0.000000	0.000000	0.000000	0.000000	0.000000
5	0.083333	0.082994	0.180310	0.090278	0.000000	0.000000	0.000000
10	0.1666667	0.1635259	0.3864430	0.1944444	−0.000000	0.0000000	−0.000000
15	0.2500000	0.2378920	0.6148770	0.3125000	−0.000000	0.0000000	−0.000000
20	0.3333333	0.3009499	0.8599127	0.4444444	−0.000000	0.0000000	−0.000000
25	0.4166667	0.3461609	1.1131836	0.5902778	−0.000000	0.0000000	−0.000000
30	0.5000000	0.3658444	1.3632775	0.7500000	−0.000000	0.0000000	−0.000000
35	0.5833333	0.3517169	1.5955682	0.9236111	−0.000000	0.0000000	−0.000000
40	0.6666667	0.2957773	1.7923844	1.1111111	−0.000000	0.0000000	−0.000000
45	0.7500000	0.1915753	1.9336531	1.3125000	−0.000000	0.0000000	0.0000000
50	0.8333333	0.0358377	1.9981497	1.5277778	0.0000000	0.0000000	0.0000000
55	0.9166667	−0.169652	1.9654489	1.7569445	0.0000001	0.0000001	0.0000001
60	1.0000000	−0.416147	1.8185950	2.0000002	0.0000002	0.0000003	0.0000002

$$\left. \begin{array}{l} y_1' = y_2 \\ y_2' = -\lambda y_1 \\ y_3' = y_4 \\ y_4' = -\lambda y_3 - g \end{array} \right\}$$

initial conditions:

$$y_1(0) = 1$$
$$y_2(0) = 0$$
$$y_3(0) = 0$$
$$y_4(0) = 0$$

(9a)

algebraic equation:

$$y_1^2 + y_3^2 - 1 = 0 \tag{9b}$$

If we let $x = \sin\theta$, $y = -\cos\theta$, substitute in the problem (8) and manipulate some derivatives and algebra, we will obtain the equivalent initial value problem of an ordinary differential equation:

$$\left.\begin{array}{l} \theta''(t) = -g\sin\big(\theta(t)\big), \ t \geq 0 \\[2mm] \theta(0) = \dfrac{\pi}{2} \end{array}\right\} \tag{10}$$

From (9b) if we differentiate the equation $y_1(t)^2 + y_3(t)^2 = 1$ twice, we will obtain the equation:

$$\lambda(t) = y_2(t)^2 + y_4(t)^2 - gy_2(t). \tag{11}$$

The system (9a) with (11) becomes a pure ODE, since the function l(t) is expressed in the terms $y_2(t)$ and $y_4(t)$ explicitly.

In this experiment, we solve (9a), (9b) using Algorithm B, and provide some of the results in Table 2. System (10) and (9a) with (11) are also solved, and they provide the same results, wherein their values are used to com- pare to the results from Algorithm B, as shown in Table 2.

Table 2: Result values for x (t), y (t) and l (t) from Algorithm B, and the errors

i	t	x	y	l(t)	x ? xs	y ? ys	l ? ls
0	0.000000	1.000000	0.000000	0.0066694	0.000000	0.000000	−0.01334
5	0.083333	0.999421	−0.034020	1.2067764	0.000000	0.000000	−0.01327
10	0.166667	0.990763	−0.135608	4.3889120	0.000000	0.000000	−0.01245
15	0.250000	0.953758	−0.300577	9.4089741	−0.000000	−0.000000	−0.00924
20	0.333333	0.858149	−0.513401	15.762372	−0.000000	−0.000000	−0.001840
25	0.416667	0.674223	−0.738528	22.336944	0.000000	0.000000	0.009746
30	0.500000	0.392046	−0.919946	27.443990	0.000000	0.000000	0.021502
35	0.583333	0.039516	−0.999219	29.406564	0.000001	0.000000	0.026667
40	0.666667	−0.320679	−0.947188	27.497515	0.000001	−0.000000	0.021650
45	0.750000	−0.621722	−0.783238	22.424766	−0.000001	0.000001	0.009949
50	0.833333	−0.826959	−0.562262	15.858262	0.000001	−0.000000	−0.001681
55	0.916667	−0.939329	−0.343017	9.4922857	0.000000	−0.000000	−0.009152
60	1.0000000	−0.986140	−0.165918	4.4483962	0.000000	−0.000000	−0.012414

The value xs, ys and ls are from problem (10). Note that the value of l(t)'s are at the mid-points t + h/2.

The third problem is an index-2 Hessenberg DAE system with nonlinear differential equations and a nonlinear algebraic equation, as follows:

$$x''(t) = x(t)(4z(t)-1)+2(1-3t)y(t), \quad 0 \le t \le 1 \qquad y''(t) = 2\sin(z(t))+y(t)(4z(t)-1)$$

initial conditions, $x(0)=0, \ y(0)=1, \ x'(0)=0, \ y'(0)=0$ algebraic equation $x^2(t)+t^2y^2(t)=t^2$

Note that this problem is similar to the first problem, but there is no variable z(t) in the algebraic equation.

Using Algorithm B, this problem is solved, and provides a nice result as shown in Table 3. The exact solution is as follows:

$$z(t)=t(1-t)$$

$$x(t)=t\sin(z(t)),$$

$$y(t)=\cos(z(t)).$$

Algorithm B gives the results for x(t) and y(t), with a small error. However, the solution for z(t) has some errors less than 0.005. Some of the results are illustrated in Table 3. Note that the value of z(t)'s are at the mid- points t + h/2.

Table 3: Result values for x (t), y (t) and z (t) from Algorithm A, and the errors

i	t	x	y	z	x ? xs	y ? ys	z ? zs
0	0.0000000	0.0000000	1.0000000	.0055058	0.0000000	0.0000000	−0.0027581
5	0.0833333	0.0063659	0.9970838	0.0860924	0.0000000	−0.0000000	0.0028285
10	0.1666667	0.0230738	0.9903704	0.1416003	0.0000000	−0.0000000	−0.0027746
15	0.2500000	0.0466008	0.9824733	0.1944127	0.0000000	−0.0000000	0.0028155
20	0.3333333	0.0734659	0.9754101	0.2221708	0.0000000	−0.0000000	−0.0027597
25	0.4166667	0.1002790	0.9706071	0.2471710	0.0000000	−0.0000000	0.0027960
30	0.5000000	0.1237020	0.9689124	0.2471938	0.0000000	−0.0000000	−0.0027367
35	0.5833333	0.1403905	0.9706071	0.2443677	0.0000000	−0.0000000	0.0027705
40	0.6666667	0.1469318	0.9754101	0.2166649	0.0000000	−0.0000000	−0.0027101
45	0.7500000	0.1398025	0.9824733	0.1860087	0.0000000	−0.0000000	0.0027448
50	0.8333333	0.1153690	0.9903704	0.1305767	0.0000000	−0.0000000	−0.0026872
55	0.9166667	0.0699551	0.9970838	0.0721021	0.0000000	−0.0000000	0.0027271
60	1.0000000	0.0000001	1.0000000	−0.0110809	0.0000000	0.0000000	−0.0026781

CONCLUSION

The methods are constructed with the objective of solving DAE systems in their original forms. Both algorithms use the Runge-Kutta method as the integrator, and couple this with a method to solve the algebraic systems associated in the problem. The Newton method is used in Algorithm A for index-1 DAE, while in Algorithm B the Newton-Broyden method is needed for an index-2 DAE system. The methods can give approximate solutions for the problem very well, with only small errors. Experiments have also shown that high index DAEs are harder to solve than lower index DAEs.

ACKNOWLEDGEMENTS

The author would like to thank the referees for their comments. This work was financially supported by the Research Grant of Burapha University through National Research Council of Thailand (Grant No. 35/2556).

REFERENCES

1. Ascher, U.M. and Petzold, L.R. (1998) Computer Method for Ordinary Differential Equations and Differential-Algebraic Equations. Society for Industrial and Applied Mathematics (SIAM), Philadelphia. http://dx.doi.org/10.1137/1.9781611971392

2. Brenan, K.E., Campbell, S.L. and Petzold, L.R. (1996) Numerical Solution of Initial-Value Problems in Differential-Algebraic Equations, Revised and Corrected Reprint of 1989 Original, with an Additional Chapter and Additional References. Classic in Applied Mathematics, 14. SIAM, Philadelphia.

3. Campbell, S.L., et al. (2008) Differential-Algebraic Equations. Scholarpedia, 3, 2849. http://dx.doi.org/10.4249/scholarpedia.2849

4. Rabier, P.J. and Rheinboldt, W.C. (2002) Theoretical and Numerical Analysis of Differential-Algebraic Equations. Hand Book of Numerical Analysis, 8, 183-540. http://dx.doi.org/10.1016/S1570-8659(02)08004-3

5. Riaza, R. (2008) Differential-Algebraic System: Analytical Espects and Circuit Applications. World Scientific, Singapore City.

6. Hairer, E. and Wanner, G. (1998) Solving Ordinary Differential Equation. II. Stiff and Differential-Algebraic Problems, 2nd Edition, Springer Series in Computational Mathematics, 14. Springer-Verlag, Berlin.

7. Chompuvised, K. and Dhamacharoen, A. (2011) Solving Boundary Value Problems of Ordinary Differential Equations with Non-Separated Boundary Conditions. Applied Mathematics and Computation, 217, 10355-10360. http://dx.doi.org/10.1016/j.amc.2011.05.044

8. Dhamacharoen, A. (2014) An Efficient Hybrid Method for Solving Systems of Nonlinear Equations. Journal of Computational and Applied Mathematics, 263, 59-68. http://dx.doi.org/10.1016/j.cam.2013.12.006

9.   Dhamacharoen, A. and Chompuvised, K. (2013) An Efficient Method For Solving Multipoint Equation Boundary Value Problems. World Academy of Science, Engineering and Technology, 7, 329-333.

10.   Differential-Algebraic Equation, Wikipedia, the Free Encyclopedia.

Chapter 7

NUMERICAL METHODS FOR SOLVING TURBULENT FLOWS BY USING PARALLEL TECHNOLOGIES

Alibek Issakhov

Department Mechanics and Mathematics, al-Farabi Kazakh National University, Almaty, Kazakhstan

ABSTRACT

Parallel implementation of algorithm of numerical solution of Navier-Stokes equations for large eddy simulation (LES) of turbulence is presented in this research. The Dynamic Smagorinsky model is applied for sub-grid simulation of turbulence. The numerical algorithm was worked out using a scheme of splitting on physical parameters. At the first stage it is supposed that carrying over movement amount takes place only due to convection and diffusion. Intermediate field of velocity is determined by method of fractional steps by using Thomas algorithm (tridiaginal matrix algorithm). At the second stage found intermediate field of velocity is used for determination of the field of pressure. Three dimensional Poisson equation for the field of pressure is solved using upper relaxation method. Moreover various ways of geometrical decomposition for parallel numerical solution of three dimensional Poisson equations are investigated.

INTRODUCTION

Most flows occurring in nature and in engineering applications are turbulent. Turbulent flow is a fluid motion that possesses complex and seemingly random structure at some macroscopic scale of dynamical importance. The most important physical consequence of turbulence is the enhancement of transport processes. In turbulent flow, momentum, energy and particle transport rates greatly exceed the corresponding molecular transport rates. Turbulent flow exhibit much more small-scale structure than their non-turbulent counterparts. In fact, this small-scale structure is correlated with enhanced turbulent transport phenomena. Small-scale structure itself is evidence of enhanced transport in the sense that small scale develop from the degradation of large-scale excitations and are maintained by energy transport from one scale to another. Another important characteristic of turbulent flows is their apparent randomness and instability to small perturbations. Currently, there are three basic and commonly used approaches for simulation of turbulent flows. First approach is direct numerical simulation (DNS) which applies to solve Navier – Stokes equations, resolving all the scales of motion, with initial and boundary conditions appropriate to the considered flow. Each simulation produces a single realization of the flow. The DNS approach was infeasible until the 1970s when computers of sufficient power became available. In DNS whole range of spatial and temporal scales of the turbulence must be resolved. All the spatial scales of the turbulence must be resolved in the computational mesh, from the smallest dissipative scales (Kolmogorov microscales), up to the integral scale L, associated with the motions containing most of the kinetic energy. Second approach is large eddy simulation (LES), the larger three – dimensional unsteady turbulent motions are directly represented, whereas the effects of the smaller-scale motions are modelled. In computational expense, LES lies between Reynolds-stress models and DNS. Because the large-scale unsteady motions are represented explicitly, LES can be expected to be more accurate and reliable than Reynolds-stress models for flows in which largescale unsteadiness is significant – such as the flow over bluff bodies, which involves unsteady separation and vortex shedding. The computational cost of DNS is high, and it increases as the cube of the Reynolds number, so that DNS is inapplicable to high Reynolds number flows. Nearly all of the computational effort in DNS is expended on the smallest, dissipative motions, whereas the energy and anisotropy are contained predominantly in the larger scales of motion. In LES, the dynamics of the large-scale motions are computed explicitly, the influence of the smaller scales being represented by simple models. Third approach is the Reynolds-averaged Navier–Stokes equations (or RANS equations) are time-averaged equations of motion for fluid flow. The idea behind the equations is Reynolds decomposition, whereby an instantaneous quantity is decomposed into its time-averaged and fluctuating quantities, an idea was first proposed by Osborne Reynolds. The RANS equations are primarily used to describe turbulent flows. These equations can be used with approximations based on knowledge of the properties of flow turbulence to give approximate time-averaged solutions to the Navier–Stokes equations.

MATHEMATICAL MODEL

Under the assumption of incompressible flow, the dimensionless governing equations are as follows [1,2,7]:

$$\frac{\partial u_i}{\partial t} + \frac{\partial u_j u_i}{\partial x_j} = -\frac{\partial p}{\partial x_i} + \frac{1}{Re}\frac{\partial}{\partial x_j}\left(\frac{\partial u_i}{\partial x_j}\right) - \frac{\partial \tau_{ij}}{\partial x_j}$$

(1)

$$\frac{\partial u_j}{\partial x_j} = 0 \quad (i = 1, 2, 3).$$

(2)

where $\tau_{ij} = \overline{u_i u_j} - \overline{u_i}\,\overline{u_j}$

The solution of spread of flow in three dimensional areas were considered in this work. u_i velocity, p represents the total pressure. The Reynolds number is defined as $Re = DV/v$ (v dynamic viscosity). Furthermore Cartesian coordinate system is employed, in which z is stream wise direction, x, y are in the lateral directions.

As for constructing model of turbulence we used dynamic model of Smagorinsky, the following is the underlying principle of the dynamic model for extracting information concerning a given eddy-viscosity model via a double filtering in physical space. It is worth to admit that the most of the historical developments have been done with Smagorinsky's model [6,9]

$$\tau_{ij} - \frac{1}{3}\delta_{ij}\tau_{kk} = -2v_{sgs}\overline{S}_{ij}$$

(3)

$$\delta_{ij} = \begin{cases} 1, i = j \\ 0, i \neq j \end{cases} \text{ Kroneker symbol where } v_{sgs} = (C_s\Delta)^2\sqrt{2\overline{S}_{ij}\overline{S}_{ij}},$$

$$\overline{S}_{ij} = \frac{1}{2}\left(\frac{\partial \overline{u}_i}{\partial x_j} + \frac{\partial \overline{u}_j}{\partial x_i}\right), \Delta = (\Delta x \Delta y \Delta z)^{1/3}$$

(4)

$$C_s = \frac{1}{\pi}\left(\frac{3C_k}{2}\right)^{-3/4}, C_s = 0.18 \text{ for a Kolmogorov constant of } 1,4.$$

But the dynamic procedure applies in fact to the types of eddy viscosities such as those used in the structure-function model.

We start with regular LES corresponding to a "bar-filter" of width Δx, an operator associating an function $\overline{f}(x,t)$. Then we define a second "test filter" tilde of large width $2\Delta x$ associating $\tilde{\overline{f}}(x,t)$. So let us first apply this filter product to the Navier-Stokes equation. The subgrid-scale tensor of the field

$\tilde{\bar{u}}_i$ is obtained from equation (4) with the replacement of the filter bar by the double filter and tilde filter:

$$\tau_{ij} = \overline{\tilde{\bar{u}}_i \, \tilde{\bar{u}}_j} - \overline{\tilde{u}_i u_j}$$

(5)

$$l_{ij} = \overline{\tilde{\bar{u}}_i \, \tilde{\bar{u}}_j} - \overline{\tilde{\bar{u}}_i \, \tilde{\bar{u}}_j}$$

(6)

Now we apply the tilde filter to equation (4), which leads to

$$\tilde{\tau}_{ij} = \overline{\tilde{u}_i u_j} - \overline{\tilde{u}_i u_j}$$

(7)

Adding equations (6) and (7) and using equation (5), we obtain

$$l_{ij} = \tau_{ij} - \tilde{\tau}_{ij}$$

Further we use Smagorinsky's model expression for the subgrid stresses related to the bar filter and tilde-filter to get

$$\tilde{\tau}_{ij} - \frac{1}{3}\delta_{ij}\,\tilde{\tau}_{kk} = -2C\,\tilde{A}_{ij}$$

where
$$A_{ij} = (\Delta x)^2 \left|\overline{S}\right|\overline{S}_{ij}$$

(8)

We have to determine τ_{ij}, the stress resulting from the filter product. This is again obtained using the Smagorinsky model, which yields

$$\tau_{ij} - \frac{1}{3}\delta_{ij}\tau_{kk} = -2CB_{ij}$$ where $$B_{ij} = (2\Delta x)^2 \left|\tilde{\overline{S}}\right|\tilde{\overline{S}}_{ij}$$

(9)

Subtracting (8) from (9) with the aid of Germano's identity we get the following

$$l_{ij} - \frac{1}{3}\delta_{ij}l_{kk} = 2CB_{ij} - 2C\,\tilde{A}_{ij}$$

$$l_{ij} - \frac{1}{3}\delta_{ij}l_{kk} = 2CM_{ij}$$

where $$M_{ij} = B_{ij} - \tilde{A}_{ij}$$

(10)

All the terms of equation (10) may now be determined by means of $\overline{u}$. Unfortunately, there are five independent equations for only one variable C, and thus the problem is over determined. The first solution was proposed by Germano to multiply (10) tensorially by $\overline{S}_{ij}$ to get 1.

$$C = \frac{1}{2} \frac{l_{ij} \overline{S}_{ij}}{M_{ij} \overline{S}_{ij}}$$

NUMERICAL SIMULATION

The numerical solution of system is built on the staggered grid with the usage of the compact scheme for convective terms and scheme against a stream of the second type [5].

The scheme of splitting on physical parameters is used for the solution of turbulence problem [9-12,14]:

I. $\quad \dfrac{\vec{u}^* - \vec{u}^n}{\tau} = -\left(\vec{u}^n \nabla \vec{u}^* - \nu \Delta \vec{u}^*\right),$

II. $\quad \Delta p = \dfrac{\nabla \vec{u}^*}{\tau},$

III. $\quad \dfrac{\vec{u}^{n+1} - \vec{u}^*}{\tau} = -\nabla p.$

The first stage is solved by fractional step method in combination of Thomas algorithm (tridiaginal matrix algorithm) [8, 11,13].

Three dimensional Poisson equation for pressure field using an over - relaxation method is handled at the second stage. Three dimensional Poisson equation is parallelized by using various geometrical decomposition (1D, 2D and 3D). Geometric decomposition of the grid area is selected as the basic approach of parallelization. In this case, there are three different ways of sharing the values of the grid function on the compute nodes one-dimensional, two-dimensional and three-dimensional of the grid computing nodes [3,4].

After a stage of decomposition, when performed on separate data blocks for the construction of a parallel algorithm, we proceed to relation between the blocks, the calculations which will be run parallel. Because of we used an explicit difference scheme for computing the next approximation in the border nodes of each subdomain is necessary to know the value of the grid function with bordering neighboring processor elements. To accomplish this, in each compute node a fake edge for storing data from a neighboring computational node and arranged shipment of these boundary values needed to ensure the homogeneity of the calculations by explicit formulas. Sending data is done using the procedures library MPI. Let us turn to a preliminary theoretical analysis of the effectiveness of various methods of decomposition of the computational domain for this case. We will estimate the time of the parallel program as the time of consistent program T_{calc}, divided by the number of processors used, plus the time shipments $T_p = T_{calc}/p + T_{com}$. While shipments to different ways of decomposition can be approximately expressed in terms of the amount of bandwidth:

$$T_{com}^{1D} = t_{send} 2N^2 x2$$

$$T_{com}^{2D} = t_{send} 2N^2 x4/p^{1/2}$$

$$T_{com}^{3D} = t_{send} 2N^2 x6/p^{2/3}$$

(11)

where N^3 - dimension of finite-difference problems, p – number of computing nodes, t_{send} - shipping time of one number.

Calculations were performed on a cluster system URSA KazNU after al-Farabi on grids of 128 × 128 × 128 and 256 × 256 × 256 by using up to 64 processors. Results of computational experiment showed the presence of a good speed in solving problems of this class. They are mainly focused on over-time shipments and time calculations for various methods of decomposition.

In the first stage we used one overall program, the size of arrays from run to run have not changed, each processor element numbering of the array elements starting from scratch. Despite the fact that, in accordance with the theoretical analysis of the 3D decomposition is the best option for parallelization (**Figure 1**), computational experiments have shown that better results can be achieved using 2D decomposition when the number of processes from 25 to 144 (**Figure 2**)

On the basis of preliminary theoretical analysis of the graphs it must have the following pattern. Computation time without interprocessor communication costs at different ways of decomposition should be approximately the same for the same number of processors and shrink as T_{calc}/p. In reality, the calculated data (**Figure 4**) indicate that the use of 2D decomposition on different grids gives the minimum cost for computation and payment schedules depending on the computation time on the number of processors which placed much higher than T_{calc}/p.

To explain these results there is a need to pay attention to the assumptions that were made during the preliminary theoretical analysis of the problem.

Firstly, it was assumed that regardless of how the distribution of data on a single processor element executed the same amount of computational work, which should lead to identical time-consuming. Secondly, we assumed that the time spent on interprocessor shipping any order of the same amount of data that does not depend on their selection from memory. To understand what happens in reality, the next set of test calculations was held. To assess the consistency of first admission was considered when the program is run in a single-processor version, and thus simulates different ways of geometric data decomposition for the same amount of computation performed by each processor.

Thus, for explicit difference methods for solving three dimensional Poisson equation can be applied one-dimensional, two and three-dimensional decomposition, but the results of testing programs have shown that the 3D decomposition does not gain in time compared with the 2D decomposition, at least for the number of processors does not exceed 250, and the 3D decomposition has a more time-consuming software implementation and the use of 2D decomposition is sufficient for the scale of the problem at the present number of compute nodes.

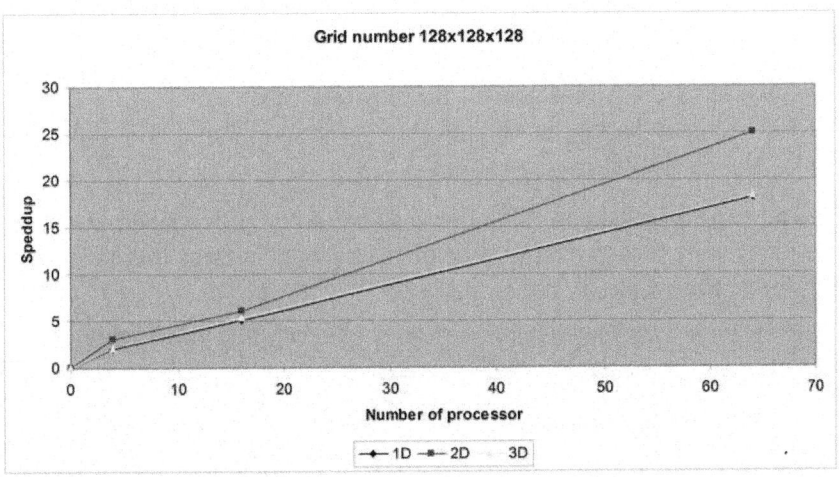

Figure 1: Speed up for different ways to decompose the computational domain

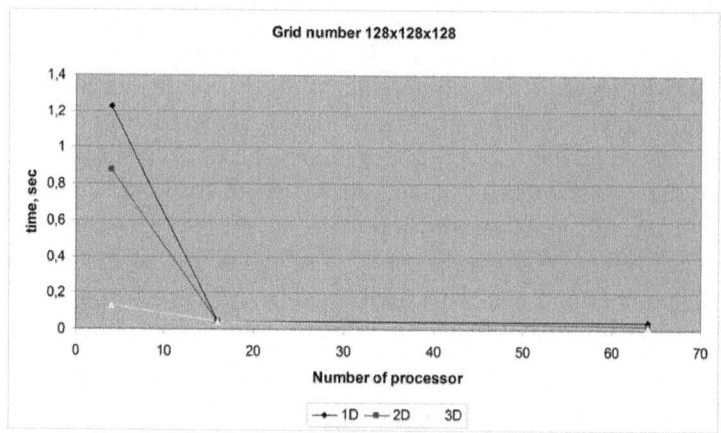

Figure 2: Computation time without considering the cost of data transfer for various methods of decomposition

TESTING RESULTS OF THE NUMERICAL METHOD

Consideration of a turbulent flow, which is located in the channel (Figure 1). Computations were performed for the Reynolds number $Re = U_m D / v$ equal to 8000 defined based on the jet axis velocity. Also the following grid $N_x x N_y x N_z = 80x80x160$ is taken in the calculations.

The spread of flow in three dimensional areas is described in numerical solution. Figure 6 shows isosurface of spread flow in three dimensional areas at different time scale.

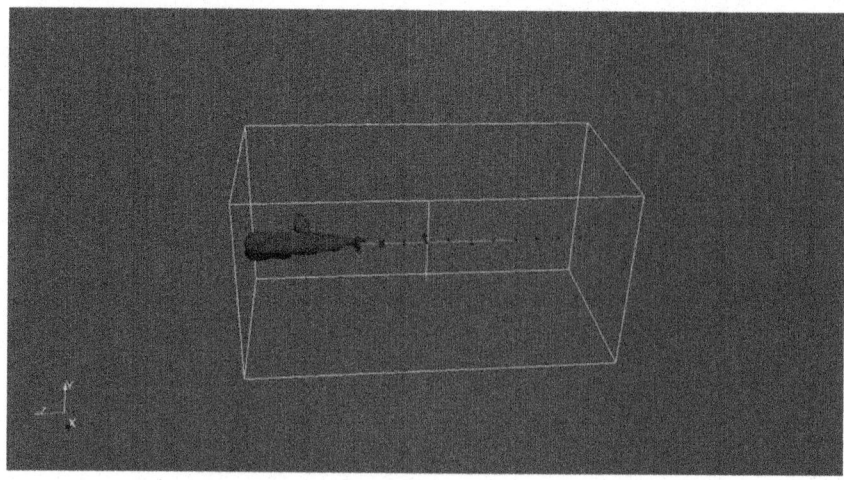

CONCLUSIONS

The results of numerical experiments showed that the constructed mathematical model of turbulence is able to reproduce the characteristic features of turbulent flow. The usage of dynamic Smagorinsky model allowed us to obtain good data for the study area. Application in the calculation of 2D decomposition gives 65% efficiency in the use of 25 compute nodes. With further increase in the number of compute nodes and 100 for the chosen mesh size, a characteristic was obtained for problems of this class efficiency value is around 45%.

REFERENCES

1. J.D. Jr. Anderson, "Computational Fluid Dynamics", New York: McGraw-Hill. 1995.

2. C.A. Fletcher, "Computational Techniques for Fluid Dynaimics," Vol 2: Special Techniques for Differential Flow Categories, Berlin: Springer-Verlag. 1988.

3. G. E. Karniadakis, "Parallel Scientific Computing in C++ and MPI." 2000

4. P. Pacheco. "Parallel Programming with MPI," Morgan Kaufmann. 1996.

5. S.K. Lely. "Compact finite difference scheme with spectral-like resolution," J. Comp. Phys., 183, 1992, pp. 16-42.

6. M. Lesieur, O. Metais, P. Comte, "Large-eddy simulations of turbulence," Cambridge university press. 2005.

7. R. Peyret, D. Th. Taylor, "Computational Methods for Fluid Flow," New York: Berlin: Springer-Verlag. 1983.

8. N.N. Yanenko, "The Method of Fractional Steps," New York: Springer-Verlag. In J.B.Bunch and D.J. Rose (eds.), Space Matrix Computations, New York: Academics Press. 1979.

9. A. Issakhov, "Large eddy simulation of turbulent mixing by using 3D decomposition method," J. Phys.: Conf. Ser. 318(4), 042051, 2011. doi: 10.1088/1742-6596/318/4/ 042051

10. B. Zhumagulov, A. Issakhov, "Parallel implementation of numerical methods for solving turbulent flows," Vestnik NEA RK. 1(43), 2012, pp. 12–24

11. A. Issakhov, "Parallel algorithm for numerical solution of three-dimensional Poisson equation," Proceedings of world academy of science, engineering and technology 64, 2012, pp. 692–694.

12. A. Issakhov, "Mathematical modeling of the influence of hydrothermal processes in the water reservoir," Proceedings of world academy of science, engineering and technology 69, 2012, pp. 632–635.

13. A. Issakhov, "Mathematical modelling of the influence of thermal power plant to the aquatic environment by using parallel technologies," AIP Conf. Proc. 1499, 2012, pp. 15-18. doi: http://dx.doi.org /10.1063/ 1.4768963

14. A. Issakhov, "Development of parallel algorithm for numerical solution of three-dimensional Poisson equation," Journal of Communication and Computer. Volume 9, Number 9, 2012, pp. 977-980.

Chapter 8

A REVIEW AND UPDATE OF ANALYTICAL AND NUMERICAL SOLUTIONS OF THE TERZAGHI ONE-DIMENSIONAL CONSOLIDATION EQUATION

Cheikhou Ndiaye[1], Meissa Fall[1], Mapathe Ndiaye[1], Daouda Sangare[2], Abib Tall[1]

[1]Laboratoire de Mécanique et Modélisation, UFR Sciences de l'Ingénieur, Université de Thiès, Thiès, Sénégal

[2]Laboratoire d'Analyse Numérique et d'Informatique, UFR Sciences Appliquées et Technologie, Université Gaston Berger, Saint-Louis, Sénégal

ABSTRACT

Practical resolution of consolidation problems that we often face requires an extensive and solid knowledge of the different parameters highlighted by the Terzaghi one-dimensional consolidation theory. This theory, with its assumptions, leads to a partial differential equation of second order in space and first order in time of pore water pressure. Analytical and numerical resolutions of this equation allow determining the water pressure variation before and after the application of a charge. Numerical modeling has enabled the simulation of the whole results obtained by the two methods of resolution (pressure, degree of consolidation, time factor, among others) to have a physical analysis and a lawful observation that lead to a suitable understanding of the phenomenon of Terzaghi one-dimensional consolidation.

INTRODUCTION

The unidimensional consolidation of soils has been described by Terzaghi using partial differential equations [1] . The resolution of these equations can be performed analytically and/or numerically [2] [3] . In this work, we resolved Terzaghi partial differential equations using Fourier method and finite difference respectively for analytical and numerical solutions. At a further step we used numerical modeling to represent graphically the two solutions in order to validate the obtained results.

We consider an example of a clay layer drained on the faces and submitted instantly to the initial conditions to constant stress (load) [4] (Figure 1). It is assumed that the clay layer thickness is 2H with H = 8 m subject to its surface to a total stress $\Delta\sigma_v$ = 30 kPa and with a consolidation coefficient estimated to 2 m²/year.

Solving the Terzaghi one-dimensional consolidation equation, from the hydro mechanical modeling of the solid [2] [3] , and adapted to the parameters of the problem allows for the following simulations:

$$\frac{\partial \Delta u(z,t)}{\partial t} = c_v \frac{\partial^2 \Delta u(z,t)}{\partial z^2}$$

(1)

where Δu is the pore water pressure, c_v is the coefficient of consolidation, z is the depth and t corresponds to the time.

ANALYTICAL METHOD

To solve this kind of first-order partial differential equation with respect to time and second-order with respect to space, we must combine two boundary conditions and initial condition for the interstitial pressure. Boundary conditions (for all time t)-At the bottom of the layer, z = 0, we have: $\Delta u(0,t) = 0$ -On the surface of the layer, z = 2H, then: $\Delta u(2H,t) = 0$ Initial condition (for t = 0) $\Delta u(z,0) = \Delta\sigma_v$ except for $z = 0$ and $z = 2H$

If we apply the Fourier sine transform F_s on the left and right members of this equation:

$$F_s\left(\frac{\partial \Delta u}{\partial t}\right) = F_s\left(c_v \frac{\partial^2 \Delta u}{\partial z^2}\right)$$

(2)

where

$$F_s\left(c_v \frac{\partial^2 \Delta u}{\partial z^2}\right) = c_v F_s\left[\frac{\partial^2 \Delta u}{\partial z^2}\right] = -c_v \frac{n\pi}{2H} F_c\left[\frac{\partial \Delta u}{\partial z}\right]$$

(3)

$$F_s\left(c_v \frac{\partial^2 \Delta u}{\partial z^2}\right) = -c_v \frac{n\pi}{2H}\left[\Delta u(2H,t)\cos n\pi - \Delta u(0,t) + \frac{n\pi}{2H} F_s\left[\Delta u(z,t)\right]\right]$$

(4)

The relation (2) is as follows:

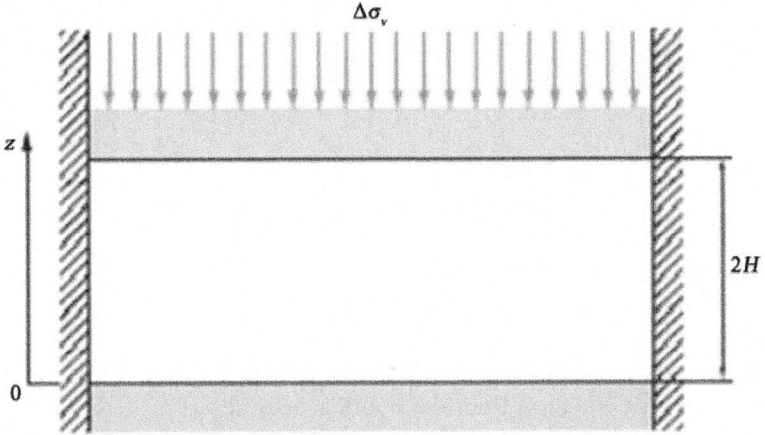

Figure 1: Example of a clay layer drained on the two faces (Modified from [4]).

$$\frac{\partial F_s\left[\Delta u(z,t)\right]}{\partial t} = -c_v\left(\frac{n\pi}{2H}\right)^2 F_s\left[\Delta u(z,t)\right] \tag{5}$$

$$\ln F_s\left[\Delta u(z,t)\right] = -c_v\left(\frac{n\pi}{2H}\right)^2 t + \ln F_s[0] \Rightarrow \ln\left(\frac{F_s\left[\Delta u(z,t)\right]}{F_s[0]}\right) = -c_v\left(\frac{n\pi}{2H}\right)^2 t \tag{6}$$

$$\frac{F_s\left[\Delta u(z,t)\right]}{F_s[0]} = e^{-c_v\left(\frac{n\pi}{2H}\right)^2 t} \Rightarrow F_s\left[\Delta u(z,t)\right] = F_s[0]e^{-c_v\left(\frac{n\pi}{2H}\right)^2 t} \tag{7}$$

Since the method used is the sine Fourier transform based on the odd frequency (sinusoidal) of the interstitial pressure, the development into Fourier series will not affect the coefficients a_n.

$$u(z,t) = \sum_{n=0}^{\infty} b_n \sin\frac{n\pi z}{2H} \tag{8}$$

where

$$b_n = \frac{2}{2H} F_s\left[\Delta u(z,t)\right]$$

By substituting, the equation can be rewritten into this form:

$$\Delta u(z,t) = \sum_{n=0}^{\infty} \frac{2}{2H} F_s\left[\Delta u(z,t)\right] \sin\frac{n\pi z}{2H} = \sum_{n=0}^{\infty} \frac{1}{H} F_s[0] e^{-c_v\left(\frac{n\pi}{2H}\right)^2 t} \sin\frac{n\pi z}{2H} \tag{9}$$

$$F_s[0] = F_s\left[\Delta u(z,0)\right] = \int_0^{2H} \Delta u(z,0) \sin\frac{n\pi z}{2H} dz = \Delta\sigma_v \int_0^{2H} \sin\frac{n\pi z}{2H} dz = -\Delta\sigma_v \frac{2H}{n\pi} \left[\cos\frac{n\pi z}{2H}\right]_0^{2H}$$

$$F_s[0] = F_s\left[\Delta u(z,0)\right] = \frac{4\Delta\sigma_v H}{(2m+1)\pi} \tag{10}$$

By simplifying and separating constants we obtain the analytical solution of the Terzaghi's equation from the Fourier method:

$$\Delta u(z,t) = \frac{4\Delta\sigma_v}{\pi} \sum_{m=0}^{\infty} \frac{1}{(2m+1)} \exp\left[-(2m+1)^2 \pi^2 \frac{c_v t}{4H^2}\right] \sin\left[\frac{(2m+1)}{2} \pi \frac{z}{H}\right] \tag{11}$$

Assuming that the $\infty = 100$, which is acceptable in numerical analysis [5], we can evaluate the numerical value of Δu (which is the exact solution) [6] [7]. For this, we make a loop for each point (z_i, t_i) and calculate $\Delta u_{exact}(z_i, t_i)$. By acting on the value t_{max} of the consolidation duration, the following results are obtained (Figure 2(a) and Figure 2(b)):

For a longer time of consolidation (e.g. 20 years), we obtain the following results (Figure 3(a) andFigure 3(b)):

For an infinite time, the phenomenon of pressure dissipation becomes more and more clear and the effective stresses are more important (Figure 4(a) and Figure 4(b)); this means that the load is transmitted to solid grains.

The degree of consolidation and the time factor are derived from the exact solution of the consolidation equation.

$$U_v(T_v) = 1 - \frac{8}{\pi^2} \sum_{m=0}^{\infty} \frac{1}{(2m+1)^2} \exp\left[-(2m+1)^2 \pi^2 \frac{T_v}{4}\right] \tag{12}$$

Hence the representation of the function $U_v(T_v)$ and the inverse function $T_v(U_v)$ gives (Figure 5): The dimensionless term which is the ratio between the value of $\Delta u(z, t)$ at (t) and its initial value $\Delta u_0 = \Delta\sigma_v$ are simulated based on the reduced depth $Z = z/H$ and for different values of T_v.

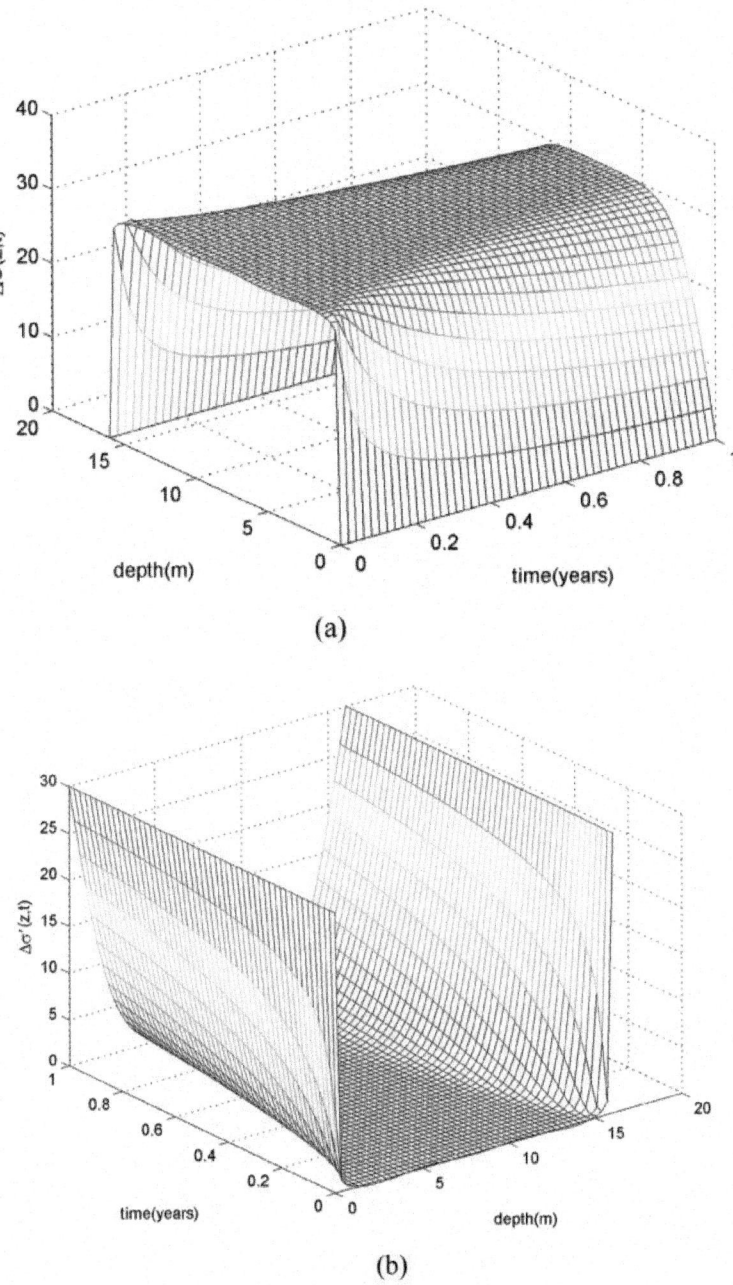

Figure 2: Evolution of the pore water pressure (a) and the effective stress (b) as a function of depth and time (we consider here t_{max} = 1 year).

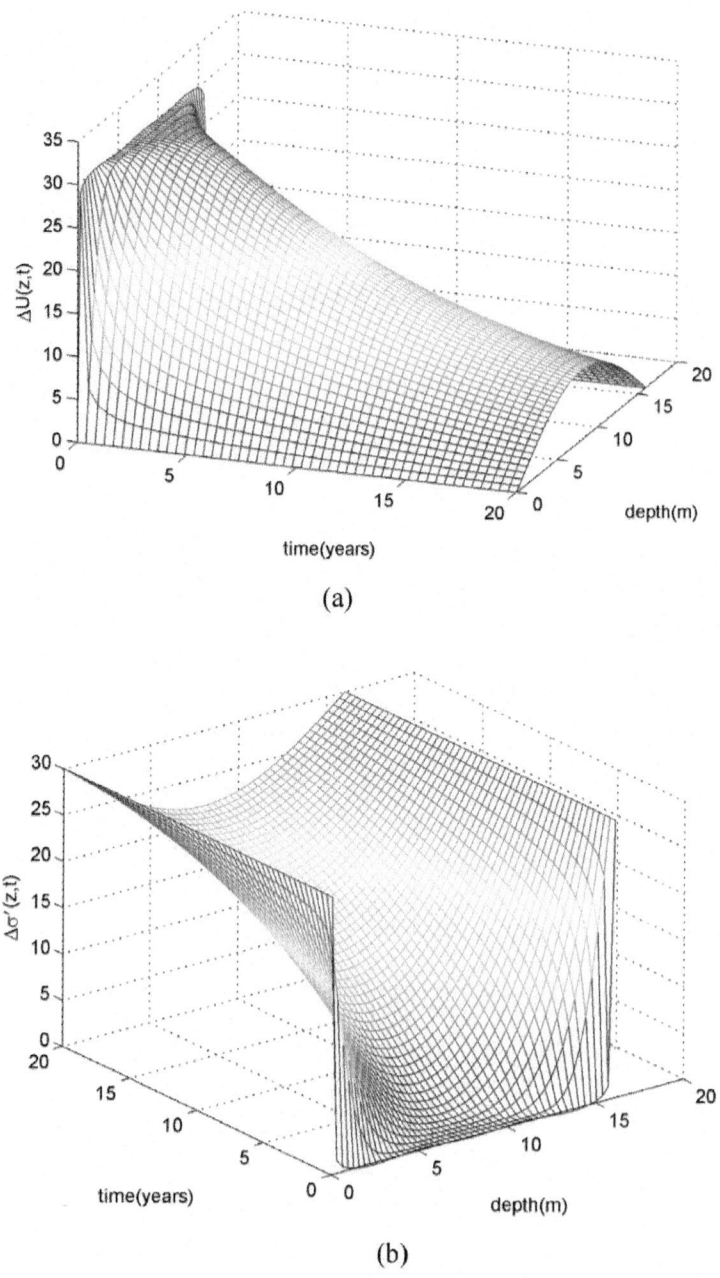

(a)

(b)

Figure 3: Evolution of the pore water pressure (a) and the effective stress (b) as a function of depth and time (we consider here t_{max} = 20 years).

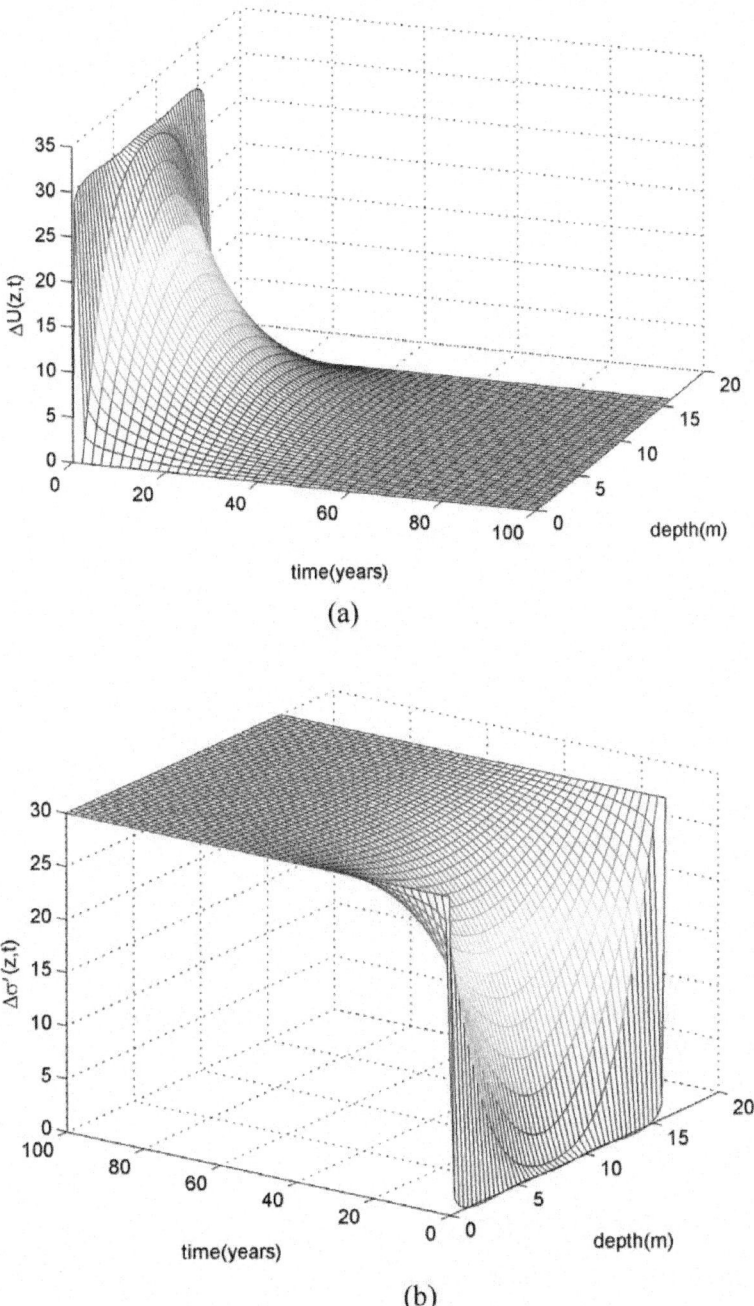

Figure 4: Evolution of the pore water pressure (a) and the effective stress (b) as a function of depth and time (we consider here t_{max} = 100 years).

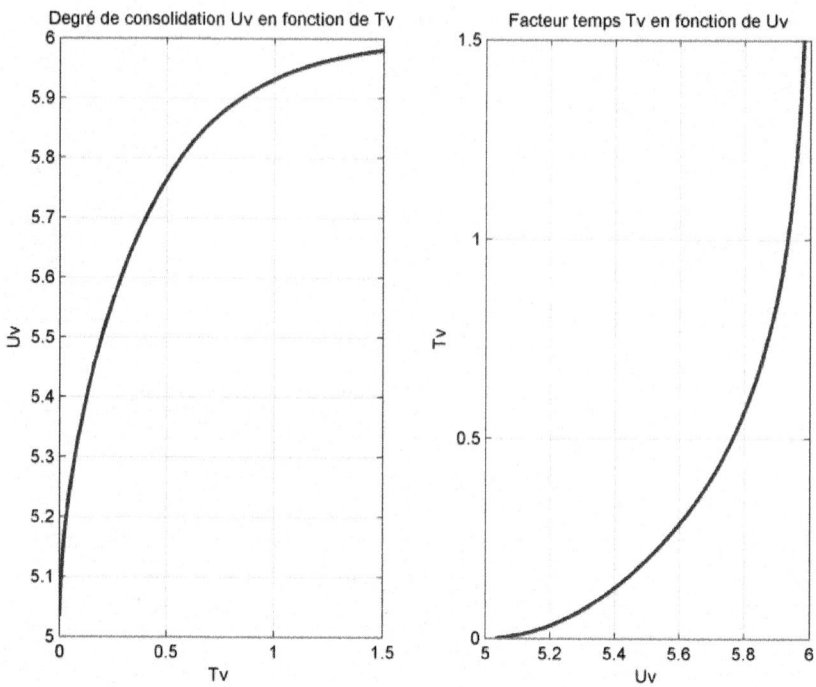

Figure 5: Respective changes of U_v and T_v, one according to others.

$$\frac{\Delta u(z,t)}{\Delta u_0} = \frac{4}{\pi} \sum_{m=0}^{\infty} \frac{1}{(2m+1)} \exp\left[-(2m+1)^2 \pi^2 \frac{T_v}{4}\right] \sin\left[\frac{(2m+1)}{2}\pi z\right]$$

(13)

Thus, we can see that the ratio $\frac{\Delta u(z,t)}{\Delta u_0}$ reaches a maximum for $T_v = 0$ and Z = 0.2 after an increasing and linear evolution for Z comprise between 0 and 0.1 (Figure 6). Then for other values of T_v with time steps ranging from 0.05 to 0.2, we obtain the other isochronous which respectively follow the first one, then with a time step of 0.1, the last isochronous; like the first isochronoous they show all ratio values that deviate more and more to 1 (which is the maximum value) when the time factor T_v tends to 1.

NUMERICAL METHOD

The principle of this method consists in substituting the function $\Delta u(z, t)$ of the interstitial pressure at the point M at time t in a discrete function $\Delta \tilde{u}(z,t)$. It requires the selection of a mesh with Δz as a space step and Δt as time step. u_{ik} or u_i^k is the interstitial pressure of water at the node (i, k). Which means that at the node $z_i = i\Delta z$ and at $t = k\Delta t$.

$$f_i' = \frac{f_{i+1} - f_i}{h} + O(h) \Rightarrow \left(\frac{\partial \Delta u}{\partial t}\right)_{ik} = \frac{\Delta u_i^{k+1} - \Delta u_i^k}{\Delta t} + O(\Delta t)$$

$$f_i'' = \frac{f_{i-1} - 2f_i + f_{i+1}}{h^2} + O(h^2) = \left(\frac{\partial^2 \Delta u}{\partial z^2}\right)_{ik} = \frac{\Delta u_{i-1}^k - 2\Delta u_i^k + \Delta u_{i+1}^k}{\Delta z^2} + O(\Delta z^2)$$

This form leads to address the resolution by an explicit scheme which uses a discretisation at z_i node and at iteration n. By analogy to the consolidation equation the equality between the first order scheme in time and the second order centered scheme in space has been set

Hence it may be evaluated to obtain:

$$\frac{\Delta u_i^{k+1} - \Delta u_i^k}{\Delta t} = c_v \frac{\Delta u_{i-1}^k - 2\Delta u_i^k + \Delta u_{i+1}^k}{\Delta z^2}$$

(14)

Given $\theta = c_v \dfrac{\Delta t}{\Delta z^2}$, this equation can be rewritten in the form that gives the interstitial pressure of the water at

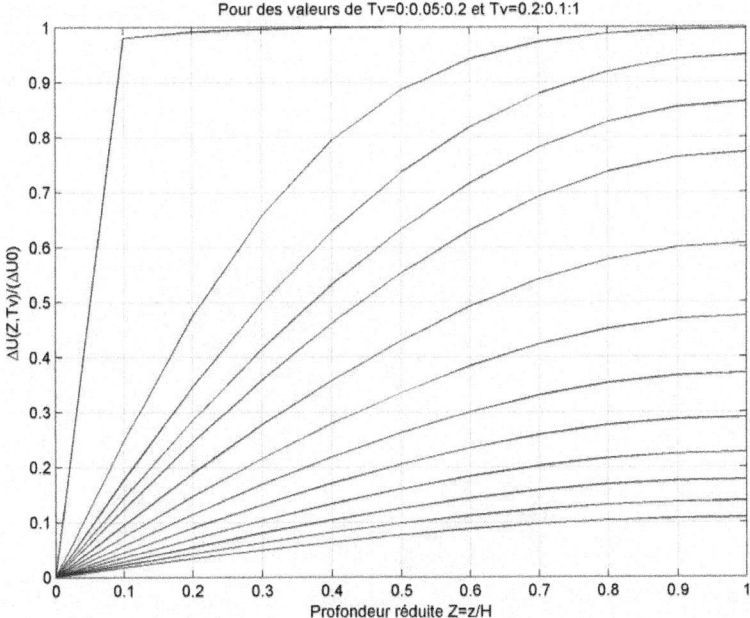

Figure 6: Isochrones of pore pressure for different values of T_v as a function of Z. iteration k +1:

$$\Delta u_i^{k+1} = \theta \Delta u_{i-1}^k + \Delta u_i^k - 2\theta \Delta u_i^k + \theta \Delta u_{i+1}^k$$

$$\Delta u_i^{k+1} = \theta \Delta u_{i-1}^k + (1 - 2\theta)\Delta u_i^k + \theta \Delta u_{i+1}^k \quad \text{where } i = 1 : N - 1 \tag{15}$$

So the matrix of excess water pore pressure from the resolution is given by the following finite differential method:

$$
\begin{bmatrix} \Delta u_1 \\ \Delta u_2 \\ \vdots \\ \Delta u_{N-2} \\ \Delta u_{N-1} \end{bmatrix}^{k+1}
=
\begin{bmatrix}
1-2\theta & \theta & 0 & \cdots & & 0 \\
\theta & 1-2\theta & \theta & \cdots & & 0 \\
\vdots & \ddots & \ddots & \ddots & & \vdots \\
0 & 0 & \theta & 1-2\theta & \theta \\
0 & 0 & 0 & \theta & 1-2\theta
\end{bmatrix}
\begin{bmatrix} \Delta u_1 \\ \Delta u_2 \\ \vdots \\ \Delta u_{N-2} \\ \Delta u_{N-1} \end{bmatrix}^{k}
+ \theta
\begin{bmatrix} \Delta u_g \\ 0 \\ \vdots \\ 0 \\ \Delta u_d \end{bmatrix}
\tag{16}
$$

Computed results in matrix form obtained by the numerical solution give also performances that converge towards an analytical solution.

For greater stability, θ must be less than 1/2 [8] ; however, we assume that θ = 1/4 and apply the other parameters of the problem for obtaining the following figures (Figure 7(a) and Figure 7(b)): For a longer time of consolidation (e.g. 20 years), we obtain the following results (Figure 8(a) andFigure 8(b)). With these results, we can notice the phenomenon of dissipation that appears in a clear way: For an infinite time, the phenomenon of pressure dissipation becomes more clear and the effective stress more important (Figure 9(a) and Figure 9(b)). In fact, the load is transmitted to solid grains.

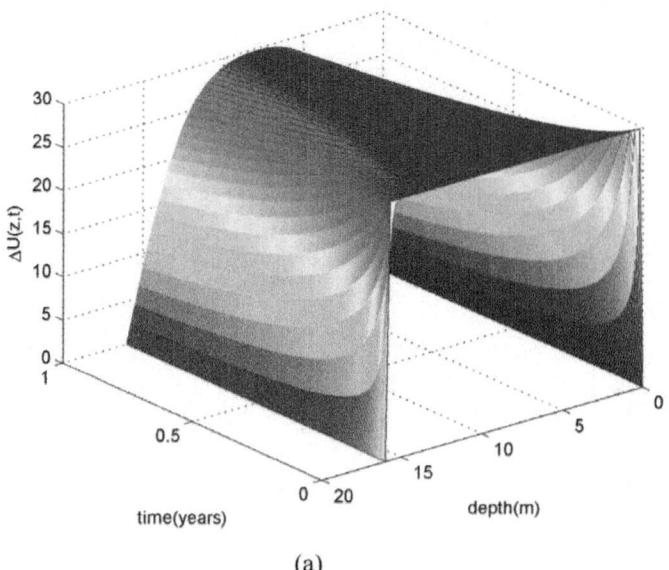

(a)

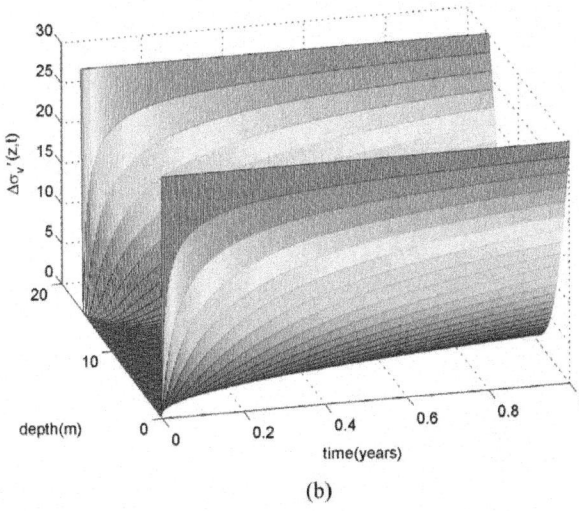

(b)

Figure 7: Evolution of the pore water pressure (a) and the effective stress (b) as a function of depth and time (we consider here tmax = 1 year).

These results show without any ambiguity that the water interstitial pressure is canceled through the soil layer thickness in the same way as noticed in the analytical resolution: The pore water pressure obviously vanishes over time with perfect coherence with Terzaghi's relation ($\Delta\sigma_v = \Delta\sigma' + \Delta u$). The load is more and more transferred towards the solid grains. Then the effective stress tends to the value of the initial stress which is applied at the soil surface.

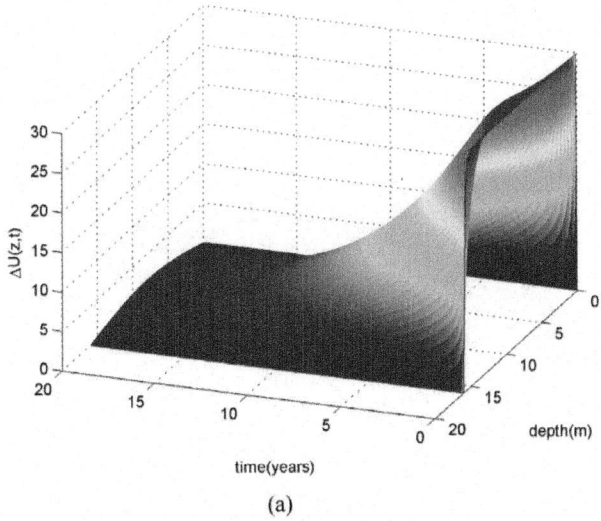

(a)

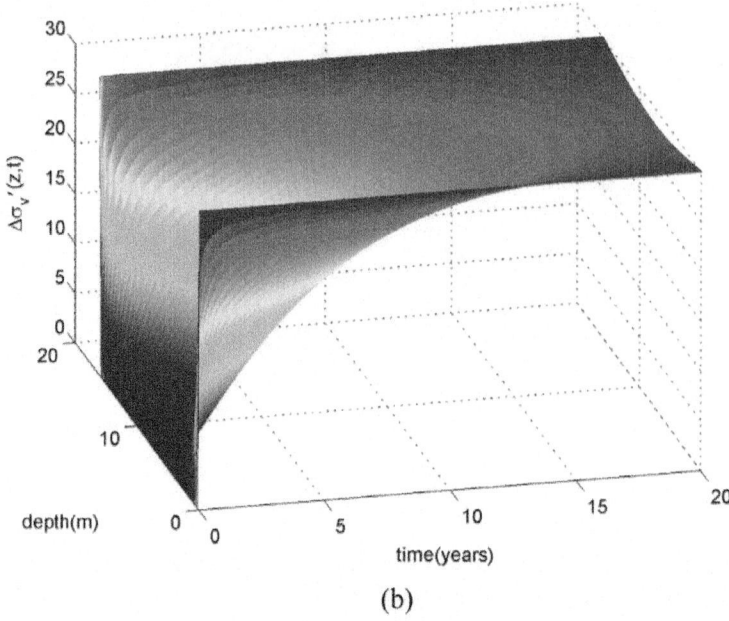

(b)

Figure 8: Evolution of the pore water pressure (a) and the effective stress (b) as a function of depth and time (we consider here t_{max} = 20 years).

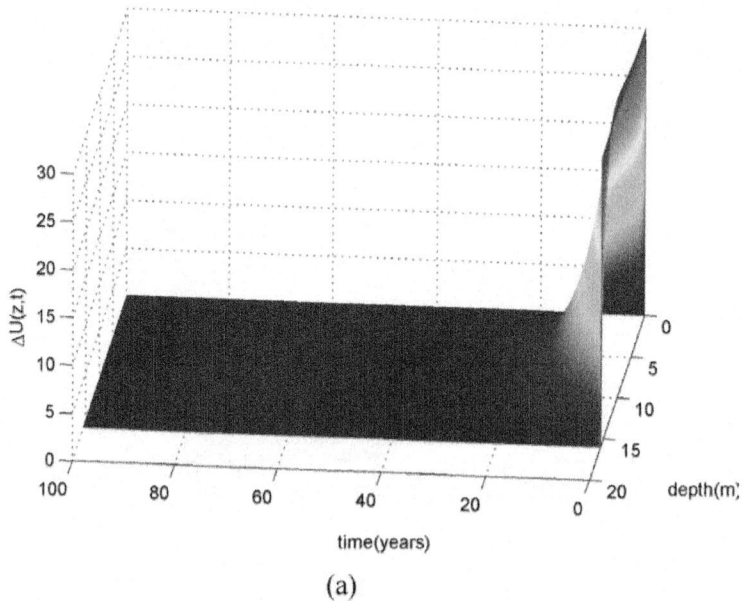

(a)

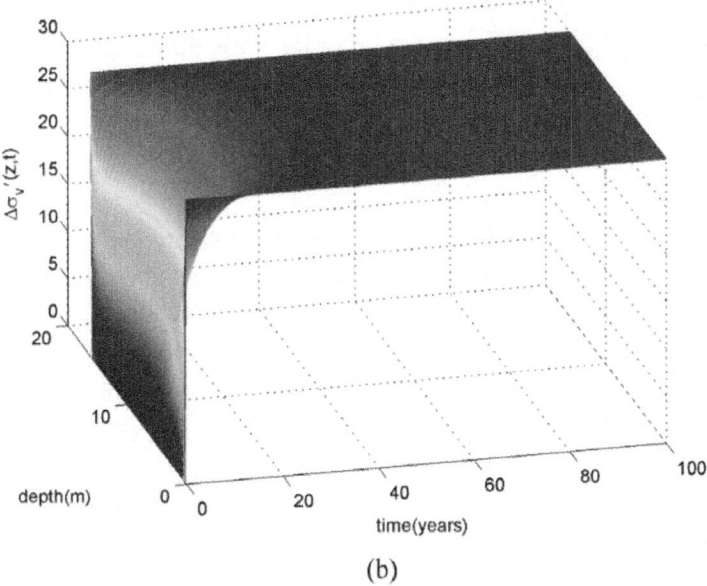

(b)

Figure 9: Evolution of the pore water pressure (a) and the effective stress (b) as a function of depth and time (we consider here t_{max} = 100 years).

CONCLUSIONS

We found for the different parameters used in our simulations that we almost have the same changes in the numerical and analytical resolutions [6] [7] . However, we notice more accuracy on the analytical resolution due to the fact that the numerical resolution gives an approximate solution while the finite differential method imposes some specific conditions that lead to a stable resolution; for the example $\theta < \dfrac{1}{2}$ for the finite differential explicit method. Nevertheless $\theta = c_v \dfrac{\Delta t}{\Delta z^2}$ implies the choice of time and space step which combination will respect the explicit scheme rule [8].

The problem of Terzaghi one-dimensional consolidation can be easily solved by analytical and numerical methods; and the solutions resulting from this resolution may also be an interesting subject for numerical simulation [9] in order to highlight more clearly the physical interpretation necessary to better understand soil consolidation.

ACKNOWLEDGEMENTS

I deeply thank the Engineering Group of the UFR Sciences de l'Ingénieur of the University of Thies and researchers from the Laboratory of Mechanics and Modeling for their guidance.

REFERENCES

1. Qin, A.F., Sun, D.A. and Tan, Y.W. (2010) Analytical Solution to One-Dimensional Consolidation in Unsaturated Soils under Loading Varying Exponentially with Time. Computers and Geotechnics, 37, 233-238.

2. Hicher, P.Y. (1985) Comportement mécanique des argiles saturées sur divers chemins de sollicitations monotones et cycliques. Application à une modélisation élastoplastique et viscoplastique. Ph.D. Thesis, Université Pierre et Marie-Curie, x p.

3. Li, X.-L. (1999) Comportement Hydromécanique des Sols Fins: De l'état saturé à l'état non saturé. Ph.D. Thesis, Sciences appliquées de l'Université de Liège, x p.

4. Magnan, J.P. and Soyez, B. (1988) Déformabilité des Sols. Consolidation. Tassement. C 214 Traité Construction, volume C 21.

5. Butcher, J.C. (1987) The Numerical Analysis of Ordinary Differential Equations: Runge-Kutta and General Linear Methods. Wiley, Wiley-Interscience.

6. The Math Works, Inc., Matlab, Reference Guide, 1984-92.

7. Torrésani, B. (2009) Introduction à Matlab et octave, Université de Province Aix Marseille I.

8. Goncalvès, E. (2005) Résolution Numérique, Discrétisation des EDP et EDO, Institut National Polytechnique de Grenoble.

9. Salazar, G.E.C. (2006) Modélisation du séchage d'un milieu poreux saturé déformable: Prise en compte de la pression du liquide. PhD thesis, Ecole Nationale Supérieure d'Arts et Métiers Centre de Bordeaux.

Chapter 9

NUMERICAL STUDY TO REPRESENT NON-ISOTHERMAL MELT-CRYSTALLIZATION KINETICS AT LASER-POWDER CLADDING

V. G. Niziev[1], F. Kh. Mirzade[1], V. Ya. Panchenko[1], M. D. Khomenko[1], R. V. Grishaev[1], S. Pityana[2], C. V. Rooyen[2]

[1]Institute on Laser and Information Technology Russian Academy of Sciences, Moscow, Russia

[2]CSIR-National Laser Centre, Pretoria, South Afric

ABSTRACT

The study of laser-powder cladding process subject to heat transfer, melting and crystallization kinetics has been carried out numerically and experimentally. The Kolmogorov-Avrami equation was applied to describe the kinetics of the phase transitions. Characteristic behavior of temperature and conversion fields has been analyzed. Melt pool dimensions, clad height dependences on mass feed rate, laser power and scanning velocity have been investigated. It has been demonstrated that the melt zone has the boundary distinct from the melting isotherm due to the fact that melting occurs with superheating and crystallization takes place at undercooling. The calculated melt pool depth and clad height are in a good agreement with the experimental results.

INTRODUCTION

Laser-powder cladding (LC) is one of the intensively developing field of laser technology in direct modification of surface characteristics of materials. It has been widely used in many practical applications such as coating, component repair, and 3D rapid prototyping. The LC needs orientation in wide range of technological regimes, search of optimal energy characteristics of laser radiation and particle flux parameters, physical properties and fraction composition of powder. The quality and microstructure of clad layer are governed by all of these parameters. Process optimization requires both theoretical and experimental understanding of the associated physical phenomena. An empirical optimization is difficult, time-consuming and expensive. Numerical modeling offers a cost-efficient way for understanding the related complex physics in a laser cladding process. An accurate model can be used in process prediction and system control.

There are several numerical and theoretical studies on melt pool dynamics in the LC process in the literature [1-4]. The 1D model of selective laser melting of twocomponents powder composition that account for heat and mass diffusion and fluid flow is investigated in [5,6]. It includes the movement of the solid particles due to shrinkage because of the density change of the powder mixture and the convective fluxes depend on surface tension and gravity forces. Liquid flow is determined by Darcy filtration law. The effect of surface settlement of the powder has been obtained.

In earlier works, height of clad layer was predefined [3]. Recently, level set method was adopted to track free surface evolution [7-10]. Han et al. [7] used a 2D mathematical model for the laser cladding to investigate powder impinging process. Qi et al. [8] developed a selfconsistent 3D model for direct metal deposition with coaxial powder injection. He et al. [9] used that model for investigation of transport phenomena. Modeling of heat-and-mass transfer during plasma deposition manufacturing has some similar solutions [10] such as level set method to track evolution of free boundary.

The knowledge of powder stream temperature is essential for laser cladding technology. Investigation of the heating and melting process in particles delivered to substrate is generally based on Stephan model [11]. Authors have modeled the dynamic and thermal behavior of powder particles in the stream by solving the coupled momentum transfer equations between the particle and gas phase. The accurate model of heating and melting of injected particles at LC process is presented in [12], where the phase change kinetics is considered. Numerical calculations and experimental measures are used [11, 13] to obtain powder stream temperature in coaxial LC.

Commonly, the kinetics of phase change is ignored both for melt pool estimations and particle melting models. In all papers mentioned above, Stephan model is used for melting and crystallization modeling reasonable for equilibrium process. However, laser material treatment is characterized as a rule by rapid and non-equilibrium processes. It is essential to account for kinetic character of the processes [14]. Multiscale model was introduced by authors [15] in order to couple microstructure evolution with macro parameters of the LC process. Nevertheless, phase change problem at LC is not studied enough.

Two approaches for modeling phase change can be underlined. The first is based on description of non-stationary nucleus evolution by unit events of atoms (molecules) attachment and detachment. The cluster size distribution function is quantity characteristic of phase change here. Its behavior is described by Fokker-Planck equation. There are different approximate solutions of it for various simplifying conditions [16-17].

The second simplified approach is in use of Kolmogorov-Avrami equation which allows to obtain explicit expression for volume fraction of new phase with the known nucleation and growth rates. This approximate approach is convenient when associated heat and phase transition kinetics equations are solved together. Earlier, this approach was used, for example, for phase change analysis of silicon exposed short laser impulses [18].

In this study, an improved phase change problem is introduced to involve rapid melting and crystallization which is typical for LC technology. The model is developed to reveal almost all phenomena occur at LC including heat diffusion, melting, crystallization, particle heat and mass addition, laser substrate interaction and free boundary evolution. It consists of 3D heat and phase change (melting/crystallization) equations solved in fully explicit way. Phase change is considered to be non-equilibrium kinetic process and associated with appearance and growth of nucleus in metastable (overheated or supercooled) medium. Modeling of phase change kinetics is based on Kolmogorov-Avrami equation.

MODEL ASSUMPTIONS

The process of powder deposition at LC is realized by feeding the powder through nozzles (often coaxial) to laser exposed area of substrate. The scheme of LC process with coaxial powder injection is presented on **Figure 1**. The substrate is exposed by laser radiation moving in positive x direction with constant velocity V. Powder of radius r_{p0} and temperature T_0 is delivered onto the substrate coaxially with the laser beam. The laser beam, when passing through the coaxial powder stream, loses energy due to reflection and absorption on particles.

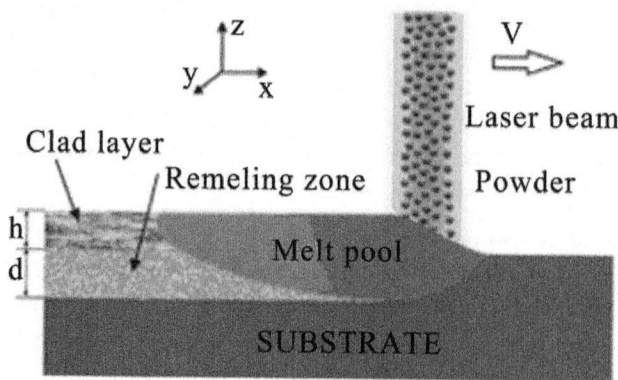

Figure 1: The scheme of laser cladding by powder injection.

Area of substrate that absorbed enough energy forms melting pool. Cooling, further crystallization and clad building up follows after, when laser beam leaves the area. Laser beam profile is assumed to be Gaussian, the depth-of-focus effect is ignored. Particles reaching molten pool get melt immediately. This approximation is made due to the fact that melting can be completed during 0.1 ms approximately for a 25 μm radius particle gained by calculation. We restrict laser intensity (I_0) to the values which does not result in particle and substrate vaporization. The higher intensity results in plasma torch formation and decreases powder capture efficiency that influences negatively on laser cladding. The effect of fluid motion due to Marangoni forces is modeled by introducing an enhanced thermal conductivity factor in the melt pool region [19].

THE PHYSICAL MODEL

The mathematical model involves heat transfer for three phases (solid, liquid and gas) with simultaneous melting and crystallization and evolution of free boundary between liquid and gas.

The Free Boundary Equation

One of the main problems at LC modeling is to obtain boundary of clad layer. Numerical modeling with moving boundaries is complex task. In earlier papers researcher used to predefine geometry of the surface. In [3], for example, 2D thermal model with spherical geometry of cladding surface was introduced to predict thermal fields.

The convenient level-set method has recently been adopted in the deposition modeling to investigate free surface evolution [2,4,7-10]. The key idea here is not to find the boundary, but to track it with the use of special function of the form:

$\phi(x,y,z,t) = \pm d$ here d is actual distance from the metal boundary, and the sign depends on the point position (in metal or gas). The function has a value in every point of the calculation domain and movement of free boundary is tracked by level set functions which satisfy the equation. The zero level set is the position of the free surface and other levels has no physical meaning. The interface tracking problem reduces to solving a partial differential equation [20]. In general case, when evolution of free boundary is governed by powder injection and hydrodynamic fluid flow, level set equation takes the form:

$$\partial\phi/\partial t + F|\nabla\varphi| = 0 \tag{1}$$

where F is the speed function that acts in the normal direction of the interface:

$$F = F_p + \mathbf{n}\cdot U(x,y,z,t),$$

$$F_p = \alpha_p \frac{8\dot{m}}{\pi\rho_p d_j^2}\exp\left(-8r^2/d_j^2\right)$$ here F_p describes the movement of the interface due to powder addition, $\mathbf{n}$ is the normal of the interface, U is the local fluid velocity α_p is the powder catchment efficiency, $\dot{m}$ is the mass feed rate, ρ_p is the powder density, d_j is the jet diameter.

The Thermal Balance equation

For two phase solid-liquid system heat transfer equation with phase change takes the form

$$cp\,\partial T/\partial t = \nabla(\lambda\,divT) + \rho(1-f_l)\Delta c\,\partial T/\partial t$$
$$- \rho L\,\partial f_l/\partial t \tag{2}$$

where c, λ и ρ are specific heat, heat conductivity and density, respectively, L is the latent heat. The detailed justification can be taken from the work [21].

In Equation (2) the first term in the right side is the heat diffusion due to thermal conductivity, the second and third terms are the power density of thermal sources due to phase change. $f_l(x,y,z,t)$ is the mass fraction of liquid phase, at the moment t ($f_l = 0$ for solid region, $f_l = 1$ is for the liquid and can change from 0 to 1 in binary region).

Laser radiation (q_{las}) is absorbed on metal-gas boundary, particles bring/take (q_p) energy depending on the particle temperature. From the other side evaporation (q_{ev}), Stephan cooling (q_{st}) occurs on the metal surface. Gas convective cooling also plays the role at LC. Diffusion of gas phase is incorporated into heat equation, so diffusion cooling is considered automatically. So for the flux on metal-gas boundary we have:

$$q_{las} = \frac{2P}{\pi R_b^2} \exp\left(-\frac{2r^2}{R_b^2}\right) \exp\left(-\frac{3Q_{ext}\dot{m}h}{\pi \rho r_p D^2 v_p}\right),$$

$$q = q_{las} - q_{con} - q_{st} = q_{las} - h(T - T_0) - \sigma \varepsilon \left(T^4 - T_0^4\right)$$

where P is the laser power, $r = \sqrt{(x - Vt)^2 - y^2}$, V is the scanning speed, second exponent in the formula for q_{las} takes in account attenuation of laser beam coaxial powder stream, r_p and v_p are the powder radius and speed, respectively, D is the diameter of powder flux on the substrate, Q_{ext} is the contact ratio. Laser beam radius is supposed to be close to powder stream radius $(Q_{ext} \approx 1)$.

Boundary condition can be included into heat transfer Equation (2) as a volume source term, using a delta function $\delta(\phi)$, where ϕ is the level-set value

$$c\rho \partial T / \partial t = \nabla \left(\lambda \, divT\right) + q\delta(\phi)$$
$$+\rho\left(1 - f_l\right)\Delta c \, \partial T / \partial t - \rho L \partial f_l / \partial t.$$

$$(3)$$

Delta function $\delta(\phi)$ for laser source incorporation can be calculated as a derivative of Heaviside function.

Coefficients for heat transfer equation are defined as follows. All coefficients in gas and metal phases are calculated using Heaviside function:

$$a = \left(1 - H(\phi)\right)a_m + H(\phi)a_g$$

here $a = (\rho, c, k)$, values with subscripts m and g denote metal and gas phase, respectively.

In this expression we use smeared Heaviside function, assigned by equation [20]:

$$H(\phi) = \begin{cases} 0, & \phi < -\varepsilon, \\ 0.5\left[1 + \phi\varepsilon^{-1} + \pi^{-1}\sin\left(\pi\phi/\varepsilon\right)\right], & |\phi| < \varepsilon, \\ 1, & \phi > \varepsilon \end{cases}$$

where ϕ is level set function described in Section 2.1, ε is half-thickness of transition region.

We suppose metal to be a continuum media which is consist of solid, liquid and mushy region. Mass fractions of liquid f_l and solid f_s phases are defined in the following

$$f_l = \left(1 + \rho_s g_s / \rho_l g_l\right)^{-1}, \quad f_s = 1 - f_l$$ where g_l and g_s are the volume fractions of liquid and solid phases, ρ_s and ρ_l are the densities of solid and liquid phases, respectively. Coefficients in mushy region are expressed by the values of mass (f) and volume (g) fractions:

$$c_m = \left(1 - f_l\right)\Delta c + c_l, \quad \Delta c = c_s - c_l,$$

$$\rho_m = \rho_s g_s + \rho_l g_l, \quad k_m = \left(\left(1 - f_l\right)k_s\rho_s^{-1} + f_l k_l \rho_l^{-1}\right)\rho_m$$

where the values with subscripts l and s denote liquid and solid phase, respectively. Using the defined mixture variables, the energy equation is applied for the whole calculation domain.

The Phase Change Kinetics

Volume fraction of new phase depends on microstructure evolution. Phase change is taken as kinetic process which is associated with nucleation and growth of nuclei of liquid/solid phase in metastable condensed matter. Let $J(T)$ and $U(T)$ are nucleation and growth rates, respectively, which are local functions of temperature. According to [22] the basic phase change kinetic law is Kolmogorov-Avrami equation, which couple volume fraction with nucleation rate (J)

$$g(r,t) = 1 - \exp\left[-\int_0^t J(r,\xi)V(r,t-\xi)d\xi\right],$$
(4)

where $V(r,t-\xi)$ is the volume of nucleus, generated to time t, and $r = (x,y,z)$. Nuclei volume estimation in single-component system is based on accounting of atoms (molecules) thermal oscillations, i.e. is determined by their transition probability through the phase boundary. In the case of spherical geometry for the nuclei volume we have $V(r,t-\xi)=(4\pi/3)\left(R(r,t-\xi)+r_*\right)^3$ where r_* the critical radius of nuclei is. Radius $R(r,t-\xi)$ is connected with the linear growth rate $U(T)$ by the following relationship

$$R=\int_{\xi}^{t} U\left(\Delta T\left(r,t''\right)\right) dt''$$

Thus kinetic Equation (5) we can rewrite in the form

$$
\begin{aligned}
&g(r,t) \\
&=1-\exp\left\{-\frac{4\pi}{3}\right. \\
&\quad \left. \times\int_{0}^{t} J\left(\Delta T(r,\xi)\right)\left[r_* +\int_{\xi}^{t} U\left(\Delta T(r,t'')\right)dt''\right]^3 d\xi\right\}
\end{aligned}
\tag{5}
$$

where $\Delta T = T - T_m$ is superheating in the case of melting, and is supercooling in the case of crystallization.

The form of the nucleation rate $J(\Delta T)$ depends on the phase change mechanism. The nucleation rate rises up rapidly on heterogeneous centers (impurities, grain boundary). But the influence of heterogeneous nucleation decreases with cooling/heating rate growth, and homogeneous nucleation plays the essential role, namely fluctuation nucleation of new phase in the unblended regions [22].

In the case then melting/crystallization occurs by homogeneous nucleation in metastable system (superheated or undercooled) the nucleation rate is described by the formula:

$$
\begin{aligned}
&J(T) \\
&= N_0\left(1-g\right)\frac{k_B T}{h}\exp\left(\frac{-E_a}{k_B T}\right)\exp\left\{\frac{-16\pi\gamma^3\Omega^2}{3k_B T\left(\Delta S\Delta T\right)^2}\right\}
\end{aligned}
\tag{6}
$$

where $N_0 = K/\Omega$ —number of atoms per unit volume (K—number of atoms per one cell, in the case of volume-centric cube K = 2; $\Omega = d_0^3$ —elementary volume of new phase, d_0 is the lattice parameter); k_B is the Boltzmann constant; E_a is the activation energy of atoms for the phase boundary transition; h is the Planck constant, γ is the surface tension; ΔS is the entropy change.

The growth rate as a function of temperature has the form:

$$U(T) = d_0 \frac{k_B T}{h} \exp\left(\frac{-E_a}{k_B T}\right)\left[1 - \exp\left(\frac{-\Delta S \Delta T}{k_B T}\right)\right]$$

(7)

The system (1), (4) and (5) is closed and can fully describe the behavior of temperature and new phase volume fraction fields at LC process with coaxial powder feeding.

NUMERICAL ALGORITHM

The numerical implementation of the heat equation is based on the semi-implicit finite-difference approximation method. The stabilizing correction method is used to solve numerical equations which are second order accuracy for time and space dimensions. The detailed procedure of calculation method can be depicted through the following iterative steps:

1)   Initialize the calculation variables: let the level set function be the signed normal distance to the surface; set liquid volume fraction g identically zero in the whole calculation domain, and T as room temperature.

2)   Compute the laser source term and the thermal coefficients for the computation domain by the temperature on previous layer.

3)   Use stabilizing correction method to solve the heat transfer equation 4) Update old liquid volume fraction with Equation (6) for new temperature.

5)   Go back to step (2) for the next iteration until convergence criterion $\sigma < \sqrt{\left(T^{i+1} - T^i\right)^2 / N}$ is reached.

6)   Advance the level set function in time in an explicit way.

7)   Update old temperature, liquid volume fraction, and level set function by new values, and go to the next time step.

NUMERICAL RESULTS AND COMPARISON WITH EX-PERIMENT

The numerical modeling of heat transfer with phase changes (melting/crystallization) at LC of Fe powder particles has been carried out using the model (1), (4) and (6). Laser power, scanning speed and powder feed rate has been varying in various regimes. Calculations have been held on 2D and 3D formulations. Single track is modeled till temperature get stationary. The convection cooling boundary condition was applied to each side of the substrate except for the free surface where clad forms. Substrate material is considered to be the same as in powder particles. Laser power varied in (0.6 -

4.4) kW diapason. Scanning speed is ranged from 6 mm/s to 4 sm/s and laser beam diameter was 0.35 - 4 mm. Powder feed rate varied in the range of 1 - 8.9 g/min and powder flow diameter was 0.8 - 2 mm on the substrate. Typical values of thermo-physical parameters for Metco 42C powder have been used in calculations.

The calculated sequential 3D track evolution is illustrated in **Figure 2**. The spatial selectivity of laser energy at LC can be seen. A small zone is affected and the material is not degraded elsewhere. The regimes where the laser heat causes substrate melting, are taken in to consideration, because only that generates a good bonding between the cladding layer and the substrate.

The cross section of temperature field along the scanning direction is presented on **Figure 3**. Temperature field is displayed by color contours and volume fraction of liquid is shown by black lines. Note that temperature profiles are stretched along the x-axis due to the influence of laser scanning speed. Melting temperature isotherm 1809K is also shown on **Figure 3** and melting/ crystallization are seen to take place with considerable superheating/ undercooling.

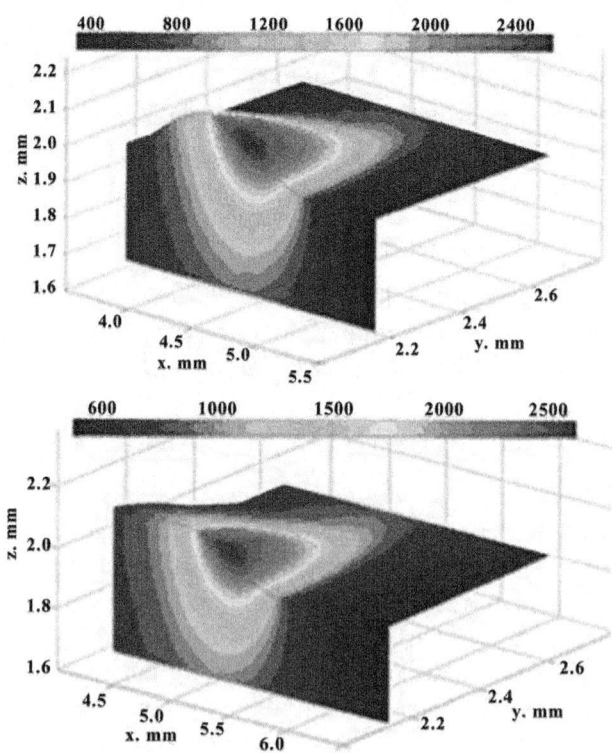

Figure 2: 3D dynamics of temperature fields; t = 62 ms (up) and t = 122 ms (down).

This occurs due to the fact that nucleation rate is zero at melting temperature, as it will be shown below, and gets significant values only after superheating (or undercooling in the case of crystallization).

The distribution of cooling rate values is introduced on **Figure 4** in x-z plane. One can note that cooling rate decreases with temperature decrease away from the laser source. Latent heat of phase change is consumed in melting region and is released in crystallization region, so cooling speed is lower in the latter one. The top of crystallization zone is cooling strongly than the bottom, because the intensive heat exchange with air.

Numerical modeling shows that cooling rate is a function of laser power and has a size of order 10^3- 10^4 K/s. The warmer crystallization zone begins to cool rapidly (cooling rate rises). Also cooling rate showed the decrease with time till temperature gets stationary.

Nucleation rate is displayed on **Figure 5** in cross section along the scanning direction by the shades of gray (with corresponding scale of values). Black dashed line represents the substrate level, and solid lines are three contours of volume fraction $g_l = 0.01$, $g_l = 0.95$ and

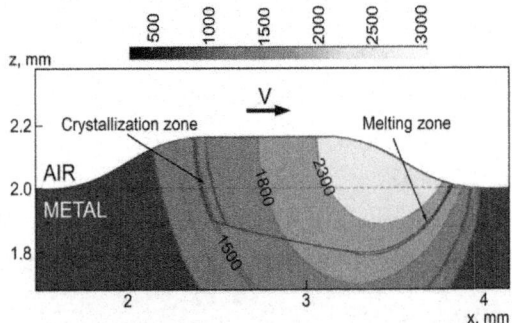

Figure 3: The x-z cross-section of temperature field, t = 130 m/s.

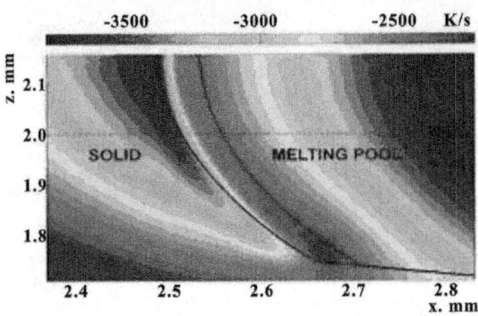

Figure 4: Spatial distribution of cooling rate in the crystallization zone P = 0.75 kW, t = 200 ms, V = 12 mm/s, m = 1 g/min.

$g_i = 0.99$. Nucleation rate is seen to get appreciable values for phase change in transition liquid-solid zone. Temperature dependences of nucleation and growth rates on the line A-A (see **Figure 5**) are presented on **Figure 6**. Note that intensive nucleation occurs at temperature. Nucleation rate increases exponentially to the values of 10^5 sm^{-3}·s^{-1}. Large quantity of nucleus restrain nucleationrate, so further growth of its value changes by rapid decrease due to solid fraction increase.

Large quantity of nucleus restrain nucleation rate, so further growth of its value changes by rapid decrease due to solid fraction increase. The calculation shows that typical undercooling [23] is reached at cooling rate observed at LC. One can note that at such undercooling first exponent in (7) associated with activation energy do not plays any role and decrease of nucleation rate occurs due to multiplier $(1-g)$. Also one can note that maximum of growth rate is in the region where nucleation doesn't occur yet, and it appears to be decreasing function in the nucleation region. Equation (6) shows that volume fraction of new phase can increase due to nucleation and growth processes.

The temperature and volume fraction dynamics are found to have qualitative difference at crystallization for different cooling rate (Figures 7 and 8). At the beginning of the process when cooling rate is high temperature change is monotonous (black solid line). Volume fraction of solid phase grows rapidly (grey solid line).

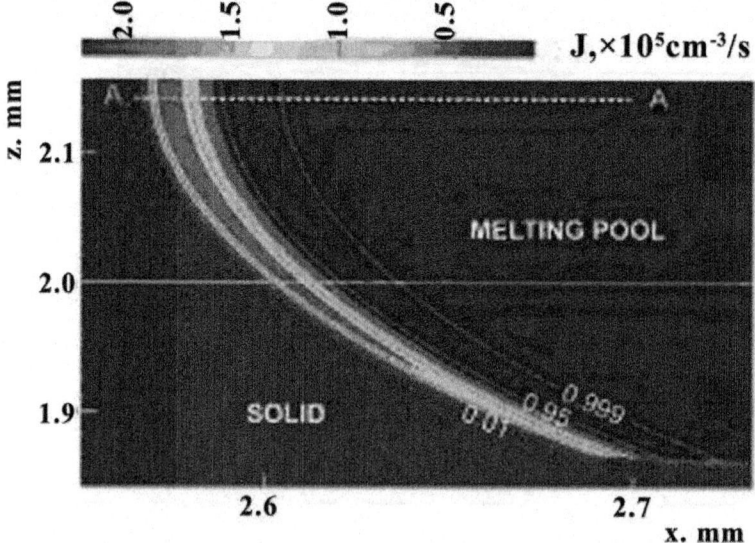

Figure 5: The x-z cross-section distribution of nucleation rate, t = 156 ms.

When the cooling rate decreases behavior of temperature is non-monotonous (black dot-dashed line) and volume fraction of solid phase grows smoother (grey, dot-dashed The laser power, scanning speed and mass feed rate are known to be the main input parameters. The output parameters are system maximum temperature, melt pool depth and cladding height. These are tree essential parameters to characterize melt pool thermal behavior and they are calculated different input parameters. The parametric investigation was conducted on 3D model. Calculated (solid line) and experimental values (dot line) of clad height and melt pool depth as functions of laser power are shown on Figures 9 and 10. The calculations are carried out at a constant mass feed rate and a scanning velocity. Melt pool size is defined by full molten state.

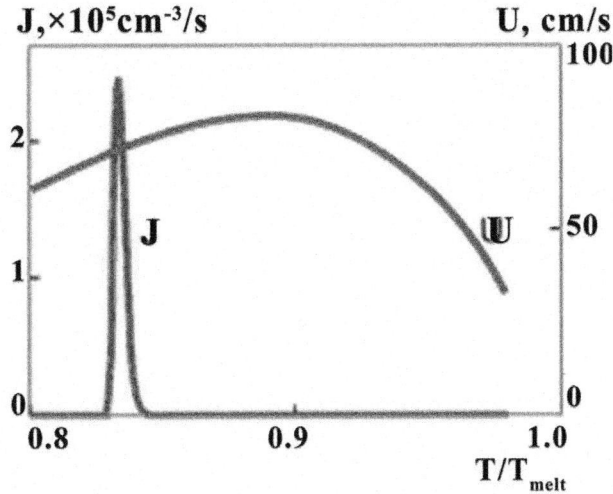

Figure 6: The temperature dependences of nucleation (J) and growth rates (v); t = 156 ms.

For comparison with the calculated clad heights were performed experiments were carried out.

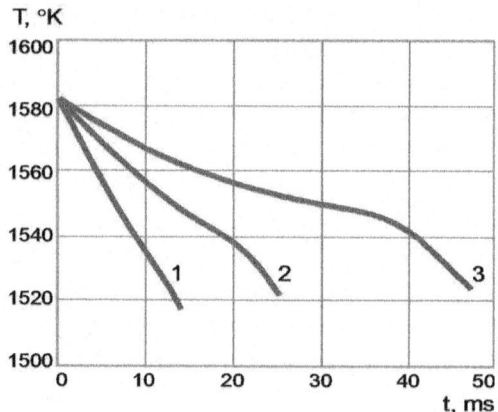

Figure 7: The temperature dynamics for different cooling rates. (1) 4500 K/s; (2) 2400 K/s; and (3) 1200 K/s.

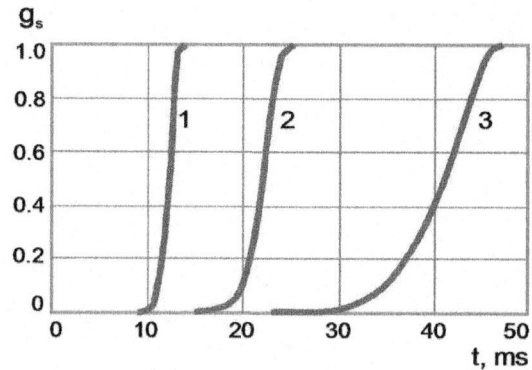

Figure 8: The time dependences of volume fraction of solid phase for different cooling rates. (1) 4500 K/s; (2) 2400 K/s, and (3) 1200 K/s.

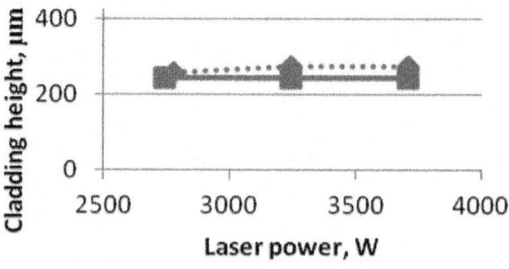

Figure 9: Clad height dependence on laser power. Dotted curve is experiment, solid line is numerical calculations. feed rate m = 8.9 g/min, scanning velocity V = 40 mm/min.

Laser clad-ding of 431 martensitic stainless steel was deposited onto 10mm flat bar C-steel substrate. A 4.4 kW Rofin DY044 diode pumped Nd: YAG laser coupled to a Kuka KR60L30 articulated arm robot and Precitec YW50 welding head with 300 mm focal length was used. Powder cladding was performed with Metco 42C powder and a coaxial nozzle with powder focus diameter of 2.0 mm. Laser spot sizes of 2.0 mm and 4.0 mm were selected to determine the effect of laser power, cladding speed and powder feed rate on dilution and clad height. The experimental methodology was based on a constant powder feed rate per unit length. Powder feed per unit length for the 2 mm spot size was 3.8 g/m and 3.8 and 9.2 g/m for the 4 mm spot size. Process parameters are shown in **Table 1**. Powder feed rate was converted from rpm to g/min by measuring the weight of powder fed per unit time.

The results of experimental data (dotted lines) for the value of the deposited layer depending on the laser power are shown in Figures 9 and 10. It is seen that the numerical results (solid line) a satisfactory agreement with the experimental data. Clad height (H) and melt pool depth (D) as functions of scanning speed are shown on **Figure 11**. Calculations are carried out at constant mass feed rate $\left(\dot{m}=1\,\text{g/min}\right)$ and for two values of laser power P = 0.6 kW and P = 0.75 kW. One observes similar decreasing trends both for clad height and melt pool depth with the increase of laser power, but the last decreases weaker.

Table 1: Processing conditions.

Powder	Spot size (mm)	Powder spot Size (mm)
42C	4	2
Power (Laser/workpiece) (W)	Scanning velocity (mm/s)	Powder feed rate (g/min)
2.781		
3.244	40	8.9186
3.707		
Laser power (kW)	gas	Gas flow rate (L/min)
3.3		
3.85	He	5
4.4		

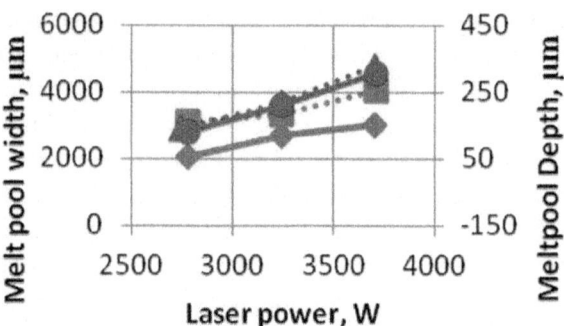

Figure 10: Experimental (dotted line) and numerical (solid line) values of melt pool width (red) and depth (blue) as a function of laser power. Feed rate m = 8.9 g/min, scanning velocity V = 40 mm/min.

The energy that is brought into melting pool increases with laser power increase which results in melt pool depth growth. The clad height influences on the melt pool dimensions so it was fixed by mass feed rate and scanning speed for melt pool depth investigations. The melt pool depth dependence on laser power at fixed clad height is presented on **Figure 12**. Results for the clad height h = 60 μm (1 g/min, V = 16 mm/s) are represented by grey line and for the clad height h = 182 μm ($\dot{m} = 3$ g/min, $V = 16$ mm/s) are shown in white color. Melt pool depth increases with laser power growth. Also one can see the influence of mass feed rate on the melt pool depth. The denser stream attenuates more laser energy so melt pool depth decreases.

CONCLUSIONS

The numerical model of LC has been presented, which considers the spatial dynamics forecasting of temperature and volume fraction of liquid phase fields at laser deposition of single-component powder on the substrates. The model is based on self-consistent nonlinear equations of heat transfer and kinetics of phase transition.

The time-space dynamics of temperature field, the free boundary evolution and profiles of distribution of liquid and solid phases for different process parameters and characteristics of laser radiation are calculated. Due to the fact that melting occurs with superheating and crystallization—with undercooling, the melt zone has the boundary distinct to the melting temperature isotherm.

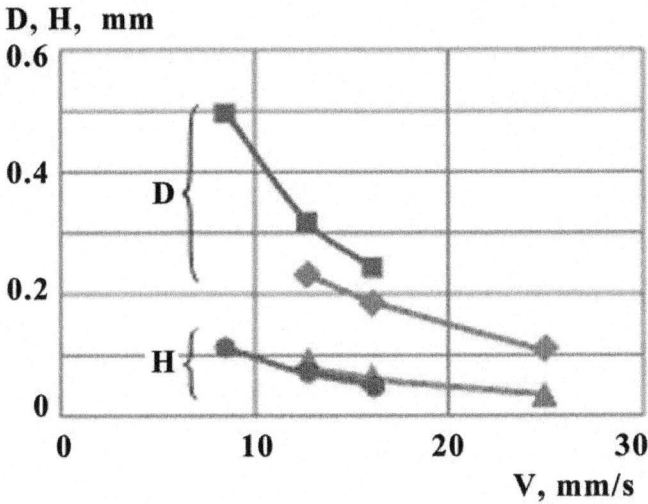

Figure 11: Melt pool depth (D) and clad height (H) dependence on scanning speed for different laser power (◆, ▲— P = 0.6 kW; ■, ●—P = 0.75 kW).

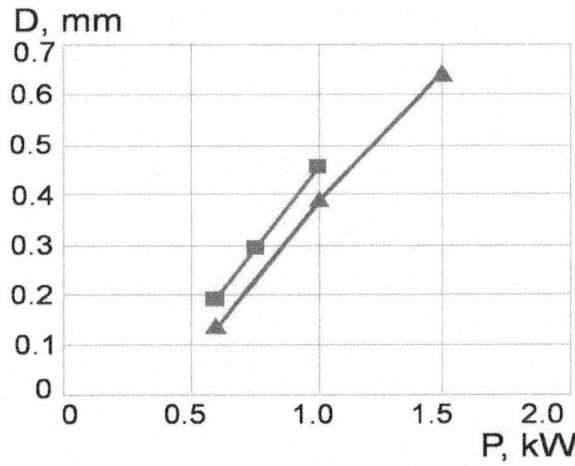

Figure 12: Melt pool depth dependence on laser power at fixed clad height: grey line (h = 60 μm, m = 1 g/min); white line—h = 182 μm, m = 3 g/min (scanning velocity V = 16 mm/s).

Qualitative difference of the crystallization process at high and low cooling speeds is discovered. Temperature dynamics is monotonous at the beginning of the process at high cooling speeds. Then, the process continues while cooling speed decreases and temperature dynamics is nonmonotonic. This is due to the influence of latent heat of crystallization: inessential of the first case and quit appreciable in the second.

The influence of laser power, mass feed rate and scanning speed on maximum temperature and the melt pool behavior is analyzed. It is determined how melt pool depth increases with laser power growth, and cladding height decreases with the scanning speed. Clad layer profile modeling agrees with experimental data. This information provides the starting data for simulations of the microstructure and the residual stress field.

ACKNOWLEDGEMENTS

This research has been supported by the Russian Foundation for Basic Research (Grant No 11-08-93966-SA_a) and Ministry of education and science of Russia (Grant No 07.524.12.4019).

REFERENCES

1. G. G. Gladush and I. Smurov, "Physics of Laser Materials Processing: Theory and Experiment," Springer-Verlag, Berlin, 2011. doi:10.1007/978-3-642-19831-1

2. S. Wen and Y. C. Shin "Modeling of Transport Phenomena during the Coaxial Laser Direct Deposition Process," Journal of Applied Physics, Vol. 108, No. 4, 2010, Article ID: 044908. doi:10.1063/1.3474655

3. A. F. A. Hoadley and M. Rappaz, "A Thermal Model of Laser Cladding by Powder Injection," Metallurgical Transactions B, Vol. 23, No. 5, 1992, pp. 631-642.

4. J. Choi, L. Han and Y. Hua, "Modeling and Experiments of Laser Cladding with Droplet Injection," Journal of Heat Transfer, Vol. 127, No. 9, 2005, pp. 978-986.

5. V. G. Niziev, A. V. Koldoba, F. Kh. Mirzade, V. Ya. Panchenko, Yu. A. Poveschenko and M. V. Popov, Mathematical Models and Computer Simulations, Vol. 3, No. 6, 2011, pp. 753-761. doi:10.1134/S2070048211060081

6. V. G. Niziev, F. Kh. Mirzade, V. Ya. Panchenko, G. V. Ustugova and V. M. Chechetkin, Mathematical Models and Computer Simulations, Vol. 3, No. 6, 2011, pp. 723-731.

7. L. Han, F. W. Liou and K. M. Phatak, "Modeling of Laser Cladding with Powder Injection," Metallurgical and Materials Transactions B, Vol. 35, No. 6, 2004, pp. 1139- 1150.

8. H. Qi, J. Mazumder and H. Ki, "Numerical Simulation of Heat Transfer and Fluid Flow in Coaxial Laser Cladding Process for Direct Metal

Deposition," Journal of Applied Physics, Vol. 100, No. 2, 2006, Article ID: 024903. doi:10.1063/1.2209807

9.   X. He and J. Mazumder, "Transport Phenomena during Direct Metal Deposition," Journal of Applied Physics, Vol. 101, No. 5, 2007, Article ID: 053113. doi:10.1063/1.2710780

10. H. O. Shang, F. R. Kong, G. L. Wang and L. F. Zeng, "Numerical Simulation of Multiphase Transient Field during plasma deposition manufacturing," Journal of Applied Physics, Vol. 100, No. 12, 2006, pp. 123522- 123531. doi:10.1063/1.2399341

11. S. Y. Wen, Y. C. Shin, J. Y. Murthy and P. E. Sojka, "Modeling of Coaxial Powder Flow for the Laser Direct Deposition Process," International Journal of Heat and Mass Transfer, Vol. 52, No. 25-26, 2009, pp. 5867-5877.

12. R. V. Grishaev, M. D. Khomenko and F. Kh. Mirzade, "Numerical Modeling of Heating and Melting of Microparticles under Laser Radiation," Proceedings of SPIE, Vol. 7994, No. 1, 2011, Article ID: 79940U1-9.

13. J. Lin, "Temperature Analysis of the Powder Streams in Coaxial Laser Cladding," Optics and Laser Technology, Vol. 31, No. 8, 1999, pp. 565-570. doi:10.1016/S0030-3992(99)00115-2

14. F. Kh. Mirzade, M. D. Khomenko, V. G. Niziev, R. V. Grishaev and V. Ya. Panchenko, "Three Dimensional Model of Melting and Crystallization Kinetics during Laser Cladding Process," SPIE Proceedings of 19th International Symposium on High-Power Laser Systems and Applications, Istanbul, 10 September 2012, Article ID: 86770R.

15. Y. Cao and J. Choia, "Multiscale Modeling of Solidification during Laser Cladding Process," Journal of Laser Applications, Vol. 18, No. 3, 2006, pp. 245-257.

16. J. W. Christian, "The Theory of Transformations in Metals and Alloys," Pergamon Press, Oxford, 1975, p. 586.

17. F. Kh. Mirzade, "Kinetics of Nucleation and Nanostructure Formation in Condensed Systems," In: V. Ya. Panchenko and V. S. Golubev, Eds., Modern Laser-Information Technologies, Intercontact Nauka, Moscow, 2005, pp. 62-78.

18. S. P. Zhvavyi, "Simulation of the Melting and Crystallization Processes in Monocrystalline Silicon Exposed to Nanosecond Laser Radiation," Technical Physics, Vol. 45, No. 8, 2000, pp. 1014-1018. doi:10.1134/1.1307010

19. C. Lampay, A. F. H. Kaplanz, J. Powellyx and C. Magnusson, "An Analytical Thermodynamic Model of Laser Welding," Journal of Physics D: Applied Physics, Vol. 30, No. 9, 1997, pp. 1293-1299. doi:10.1088/0022-3727/30/9/004

20. J. A. Sethian, "Level Set Methods and Fast Marching Methods," 2nd Edition, Cambridge University Press, Cambridge, 1999.

21. V. R. Voller and C. R. Swaminathan, "General SourceBased Method for Solidification Phase Change," Numerical Heat Transfer, Part B, Vol. 19, No. 2, 1991, pp. 175-189.

22. V. P. Skripov and V. P. Koverda, "Spontaneous Crystallization of Undercooled Liquid," Nauka, Moscow, 1984.

23. A. V. Evteev, A. T. Kosilov, E. V. Levchenko and O. B. Logachev, "Kinetics of Isothermal Nucleation in Supercooled Melt of Iron," Fizikatvyordogotela, Vol. 48, No. 5, 2006, pp. 557-582.

Chapter 10

CONVERGENCE AND ERROR OF SOME NUMERICAL METHODS FOR SOLVING A CONVECTION-DIFFUSION PROBLEM

Gabriela Nut, Ioana Chiorean, Petru Blaga

Applied Mathematics Department, Babes-Bolyai University, Cluj Napoca, Romania

ABSTRACT

We use the local Fourier analysis to determine the properties of the multigrid method when used in modeling the skin penetration of a drug. The analyses of these properties can be very in designing an efficient structure of the multigrid method and in comparing the element and finite difference discretization techniques. After the theoretical results obtained, we also present some numerical results for a problem for which the solution is known.

INTRODUCTION

In this paper we present an eoretical study of the smoothing, convergence and error reduction properties of the multigrid method for a time dependent convection diffusion equation. This is an equation that arises in the mathematical modeling of many physical phenomena, which makes the efficient numerical solution very important.

The equation studied here models the transport of molecules through the layers of the skin, until it reacheas the blood stream. The parameters used for the diffusion coefficients are smaller by several order than those of the convection, thus the equation is a convection dominated one.

The discretization of the differential equation is realized by two different methods: the finite difference method [1] with Euler backward discretization and the Galerkin finite element method [2,3].

The system obtained after the discretization process is solved using the multigrid method. This method was first introduced by Fedorenko [4,5]. The first practical results and efficiency of the method were given by Brandt [6,7].

The theory of multigrid convergence is well established for the Poisson equation [8-10]. In more recent articles the convergence has been studied for the convectiondiffusion equation [11,12].

The novelty in this paper is that we study the smoothing factor, asimptotic convergence factor and the error reduction factor of the multigrid method for a time dependent convection-diffusion equation, on a domain comprising three layers with different physical properties. The analyse is performed using the local Fourier techniques [9,13] which represents a good tool for constructing efficient multigrid methods for a given differential equation.

We also determined the error obtained for a given solution of the model problem, using the multigrid method on different number of grid levels, for both discretization methods mentioned above.

MATHEMATICAL MODEL

$$\begin{cases} c\dfrac{\partial u(x,t)}{\partial t}+v\cdot\nabla u(x,t) \\ =d\Delta u(x,t)+\alpha u(x,t)+f(x), \\ \qquad u(x_0,t)=u_0, \end{cases} \qquad \begin{aligned} &t\geq 0, x\in\Omega \\[2em] &t\geq 0 \end{aligned}$$

$u(x,t)$ represents the concentration of the substance transported through the blood stream, v is the vector of convection coefficients and d is the vector of diffusion coefficients.

The substance which is transported through the skin is applied at the surface on an area with a radius of a few centimeters. The depth to which the active substance is transported by the diffusion and convection process is of the order of nanometers, thus much smaller than the radius of the surface where it is appplied. As a consequence, the problem can be reduced to the unidimensional case. From this point on, the variable x will represent the depth where the concentration is to be calculated, and the vectors v and d are the coefficients in different layers of the skin.

The concentration of the substance applied on the skin is known, and the amount of it is sufficiently large to be constant at any moment of time t :

$$u(0,t)=u_0, t\geq 0, \tag{1}$$

this being the initial condition of the problem.

On the frontiers between the skin layers the law of flux conservation gives:

$$\left[\!\left[-d\frac{\partial u(x,t)}{\partial x}\right]\!\right] = 0, x = x_{0i}, i = 1,2,\cdots,n_d, t \geq 0,$$

$$(2)$$

n_d is the number of layers where the diffusion takes place and:

$$[\![a(x,t)]\!] = a(x^+,t) - a(x^-,t).$$

After the discretization process, the system obtained from the Equation (1) has the form:

$$\begin{cases} q_0 u_i + q_1 u_{i-1} + q_2 u_{i+1} = f_i, i = 1,2,\cdots,N, \\ u_1 = u_0, u_N = 0 \end{cases}$$

$$(3)$$

$$u_i = u(x_i, t), x_i \in G_h = \{kh, k \in \mathbb{Z}, h = (b-a)/(N+1)\}.$$

If the finite elements method [14] is used, the weak formulation of Equation (1) gives:

$$\int_0^b c\frac{du}{dt}v\,dx + \int_0^b v\frac{du}{dx}v\,dx$$

$$= \int_0^b d\frac{d^2u}{dx^2}v\,dx + \int_0^b \alpha uv\,dx + \int_0^b fv\,dx,$$

v is a function that has a derivative of order 1 and is square-integrable on $\Omega = [0,b]$. The functions u and v are approximated using the continuous functions $\Phi_i : \Phi_i(x_j) = \delta_{ij}, i,j = 1,\cdots,N$, N being the number of interior points of the grid on level l, through the relations: $u \approx \sum_{j=1}^N u_j\Phi_j, v \approx \sum_{j=1}^N v_j\Phi_i$. Replacing the functions

u and v with these approximates and using the standard integration-by-parts formula the equation becomes:

$$\sum_{i=1}^N \sum_{j=1}^N \left(\int_0^b c\Phi_j\Phi_i\frac{du_j}{dt}v_i\,dx + \int_0^b v\frac{d\Phi_j}{dx}\Phi_i u_j v_i\,dx \right.$$

$$\left. + \int_0^b d\frac{d\Phi_j}{dx}\frac{d\Phi_i}{dx}u_j v_i\,dx - \int_0^b \alpha\Phi_j\Phi_i u_j v_i\,dx \right)$$

$$= \sum_{i=1}^N \int_0^b f\Phi_i v_i\,dx$$

or:

$$\sum_{j=i-1}^{i+1} \left(c_{ij}c\frac{u_j - u_j^{ant}}{\Delta t} + vv_{ij} + dd_{ij} - \alpha a_{ij} \right) u_j = f_i,$$

$$a_{ij} = c_{ij} = \int_0^b \Phi_j \Phi_i dx; v_{ij} = \int_0^b \frac{d\Phi_j}{dx} \Phi_i dx;$$

$$d_{ij} = \int_0^b \frac{d\Phi_j}{dx}\frac{d\Phi_i}{dx} dx; f_i = \int_0^b f\Phi_i dx, i, j = 1, \cdots, N$$

(4)

Computing the integrals from (4) for

$$\Phi_i(x) = \begin{cases} \dfrac{x - x_{i-1}}{x_i - x_{i-1}}, x \in [x_{i-1}, x_i] \\ \dfrac{x_{i+1} - x}{x_{i+1} - x_i}, x \in [x_i, x_{i+1}] \\ 0, \qquad x \notin [x_{i-1}, x_{i+1}] \end{cases}$$

the coefficients in the system (3) are:

$$q_0 = -\frac{2\alpha h}{3} + \frac{2ch}{3dt} + \frac{2d}{h},$$

$$q_1 = -\frac{\alpha h}{6} + \frac{ch}{6dt} - \frac{v}{2} - \frac{d}{h},$$

$$q_2 = -\frac{\alpha h}{6} + \frac{ch}{6dt} + \frac{v}{2} - \frac{d}{h},$$

$$f_i = f(x_i) + \frac{ch}{6dt}\left(u_{i-1}^{ant} + 4u_i^{ant} + u_{i+1}^{ant} \right).$$

(5)

For the finite differences method using the explicit backward Euler scheme:

$$c\frac{u_i - u_i^{ant}}{\Delta t} + v\frac{u_{i+1} - u_{i-1}}{2h}$$

$$= d\frac{u_{i+1} - 2u_i + u_{i-1}}{h^2} + \alpha u_i + f(x_i), i = 1, \cdots, N$$

the coefficients for the system (3) will be:

$$q_0 = -\alpha + \frac{c}{dt} + \frac{2d}{h^2}, q_1 = -\frac{v}{2h} - \frac{d}{h^2},$$

$$q_2 = \frac{v}{2h} - \frac{d}{h^2}, f_i = f(x_i) + c\frac{u_i^{ant} h}{dt}.$$

(6)

In the nodes that are on the frontiers between different layers of the skin $(x_{0i}, i = 1, 2, \cdots, nd)$, the law of flux conservation (2) becomes:

$$d(x_i^-)\frac{u_{i,t} - u_{i-1,t}}{h} = d(x_i^+)\frac{u_{i+1,t} - u_{i,t}}{h},$$

$$i = 1, 2, \cdots, nd.$$

In these points the system (3) has the coefficients:

$$q_0 = \frac{1}{h}(d(x_i^-) + d(x_i^+)), q_1 = -\frac{1}{h}d(x_i^-),$$

$$q_2 = -\frac{1}{h}d(x_i^+), f_i = 0.$$

(7)

THE COMPONENTS OF THE MULTIGRID METHOD

For the components of the multigrid method we give in the following the matrices associated to their operators, needed for the local Fourier analysis of the convergence. The essential property used by this method is the fact that the discretiztion of the problem leads to a system that has the eigenvectors equals to the Fourier modes and when the multigrid components have a block structure when computed in the Fourier basis, the analysis of the multigrid method is reduced to the one of diagonal blocks of small size.

The Matrix of an Operator

If A is an operator that can be described by a difference stencil:

$$A = [\![a_1 \quad a_2 \quad a_3]\!]$$

meaning that:

$$Au(x) = a_1 u(x - h) + a_2 u(x) + a_3 u(x + h), x \in G_h,$$

then the functions $\varphi(\theta, x) = e^{\frac{i\theta x}{h}}, \theta \in [-\pi, \pi), x \in G_h$ are the eigen functions of A:

$$A\varphi(\theta, x) = \tilde{A}(\theta)\varphi(\theta, x), x \in G_h,$$

(8)

and:

$\tilde{A}(\theta) = a_1 e^{-i\theta} + a_2 + a_3 e^{i\theta}$ are the eigenvalues of A. As $\varphi(\theta, x) = \varphi(\theta + 2k\pi, x)$, it is sufficient to take $\theta \in [-\pi, \pi)$. If $\theta \in \left[-\dfrac{\pi}{2}, \dfrac{\pi}{2} \right] = T^{low}$, the set of low frequencies, then:

$\bar{\theta} = \begin{cases} \theta + \pi \text{ if } \theta < 0 \\ \theta - \pi \text{ if } \theta > 0 \end{cases} \in T^{high}$, $T^{high} = [-\pi, \pi) - T^{low}$ is the set of high frequencies.

Using the above notations, for an arbitrary function

$\Psi(x) = \alpha \varphi(\theta, x) + \beta \varphi(\bar{\theta}, x)$, the operator A applied to Ψ gives:

$$A\Psi(x) = \left(\varphi(\theta, x) \quad \varphi(\bar{\theta}, x) \right) \cdot \hat{A} \cdot \begin{pmatrix} \alpha \\ \beta \end{pmatrix}$$

and $\hat{A}$ represents the matrix associated with the operator A.

The Operator of the Discretized System

$L_h = [\![q_1 \quad q_0 \quad q_2]\!]$ from system (3) applied to a function Ψ will give:

$L_h \Psi(x) = \alpha \tilde{L}_h(\theta) \varphi(\theta, x) + \beta \tilde{L}_h(\bar{\theta}) \varphi(\bar{\theta}, x)$

$= \left(\varphi(\theta, x) \quad \varphi(\bar{\theta}, x) \right) \begin{pmatrix} \alpha \tilde{L}_h(\theta) \\ \beta \tilde{L}_h(\bar{\theta}) \end{pmatrix}$

$= \left(\varphi(\theta, x) \quad \varphi(\bar{\theta}, x) \right) \begin{pmatrix} \tilde{L}_h(\theta) & 0 \\ 0 & \tilde{L}_h(\bar{\theta}) \end{pmatrix} \begin{pmatrix} \alpha \\ \beta \end{pmatrix}$,

where $\tilde{L}(\theta) = q_0 + q_1 e^{-i\theta} + q_2 e^{i\theta}$.

Thus the matrix of L_h is:

$$\hat{L}_h = \begin{pmatrix} \tilde{L}_h(\theta) & 0 \\ 0 & \tilde{L}_h(\bar{\theta}) \end{pmatrix}. \tag{9}$$

Preand Post Smoother

The Gauss-Seidel red-black method is used before and after the coarse grid correction, and reduces well the high frequency components of the error. The smoothing operator has two components of Jacobi type:

$$S_h^{\text{red}} \varphi(\theta, x) = \begin{cases} \left(1 - \dfrac{\omega}{q_0} \tilde{L}(\theta)\right) \varphi(\theta, x), x \in G^{\text{red}} \\ \varphi(\theta, x), \qquad\qquad x \in G^{\text{black}} \end{cases}$$

$$= \frac{1}{2} \left((a+1)\varphi(\theta, x) + (a-1)\varphi(\overline{\theta}, x) \right).$$

$$S_h^{\text{black}} \varphi(\theta, x) = \begin{cases} \varphi(\theta, x), \qquad\qquad x \in G^{\text{red}} \\ \left(1 - \dfrac{\omega}{q_0} \tilde{L}(\theta)\right) \varphi(\theta, x), x \in G^{\text{black}} \end{cases}$$

$$= \frac{1}{2} \left((1+b)\varphi(\theta, x) + (1-b)\varphi(\overline{\theta}, x) \right).$$

In the relations above:

$$a = 1 - \frac{\omega}{q_0} \tilde{L}(\theta), b = 1 - \frac{\omega}{q_0} \tilde{L}(\overline{\theta}).$$

$$(10)$$

As:

$$S_h \Psi(\theta, x) = S_h^{\text{black}} S_h^{\text{red}} \Psi(\theta, x)$$

$$= \left(\varphi(\theta, x)\varphi(\overline{\theta}, x) \right) \cdot \frac{1}{4} \begin{pmatrix} 1+a & 1-b \\ 1-a & b+1 \end{pmatrix} \begin{pmatrix} a+1 & b-1 \\ a-1 & b+1 \end{pmatrix} \begin{pmatrix} \alpha \\ \beta \end{pmatrix},$$

the matrix of the smoother will be:

$$\hat{S}_h = \hat{S}_h^{\text{black}} \hat{S}_h^{\text{red}} =$$

$$\frac{1}{4} \begin{pmatrix} (a+1)^2 - (a-1)(b-1) & (b-1)(a-b) \\ -(a-1)(a-b) & (b+1)^2 - (a-1)(b-1) \end{pmatrix}$$

$$(11)$$

Restriction of the Defect

Full-weighting restriction is used as a fine to coarse grid transfer operator :

$$I_h^{2h} = \frac{1}{2} \llbracket 1 \quad 2 \quad 1 \rrbracket;$$

$$I_h^{2h} \Psi(x) = \alpha I_h^{2h} \varphi_h(\theta, x) + \beta I_h^{2h} \varphi_h(\overline{\theta}, x)$$

$$= \alpha \tilde{I}_h^{2h}(\theta) \varphi_{2h}(2\theta, x)$$

$$+ \beta \tilde{I}_h^{2h}(\theta) \varphi_{2h}(2\overline{\theta}, x).$$

As

$$\varphi_{2h}\left(2\bar{\theta},x\right)=e^{-\frac{2(\theta\pm\pi)x}{2h}}=e^{-\frac{2\theta x}{2h}}$$

$$=\varphi_{2h}\left(2\theta,x\right)\text{ for }x\in G_{2h},$$

(12)

the restriction operator applied to the function $\Psi(x),\ x\in G_h$ will give:

$$I_h^{2h}\Psi(x)=\left(\alpha\tilde{I}_h^{2h}(\theta)+\beta\tilde{I}_h^{2h}(\bar{\theta})\right)\varphi_{2h}\left(2\theta,x\right)$$

$$=\varphi_{2h}\left(2\theta,x\right)\left(\tilde{I}_h^{2h}(\theta)\ \ \tilde{I}_h^{2h}(\bar{\theta})\right)\binom{\alpha}{\beta},$$

(13)

with $\tilde{I}_h^{2h}(\theta)=\frac{1}{2}\left(e^{-i\theta}+2+e^{i\theta}\right)$.

Thus the restriction operator has the matrix:

$$\hat{I}_h^{2h}=\frac{1}{2}\left(1+\cos\theta\ \ 1+\cos\bar{\theta}\right).$$

(14)

Solution on the Coarse Grid

In the two-grid method, the exact solution on the coarse grid is required. After the restriction of the defect, the function on the grid G_{2h} has the form

$$\Psi_{2h}(x)=\alpha_{2h}\varphi_{2h}\left(2\theta,x\right)$$ (12,13). Thus:

$$L_{2h}^{-1}\Psi_{2h}=\frac{\alpha_{2h}}{\tilde{L}_{2h}(2\theta)}\varphi_{2h}\left(2\theta,x\right),$$

wherefrom the matrix of the operator is:

$$\hat{L}_{2h}^{-1}(2\theta)=\frac{1}{\tilde{L}_{2h}(2\theta)}.$$

(15)

Prolongation

The coarse to fine interpolation operator used is the bilinear interpolation:

$$I_{2h}^h\varphi_{2h}\left(2\theta,x\right)$$

$$=\begin{cases}\varphi_{2h}\left(2\theta,x\right), & \frac{x}{h}=2k\in Z,\\[2mm]\dfrac{\varphi_{2h}\left(2\theta,x+h\right)+\varphi_{2h}\left(2\theta,x-h\right)}{2}, & \frac{x}{h}=2k+1\in Z,\end{cases}$$

$$=\frac{1}{2}\left(1+\cos\theta\right)\varphi_h\left(\theta,x\right)+\frac{1}{2}\left(1+\cos\bar{\theta}\right)\varphi\left(\bar{\theta},x\right).$$

From this relation it follows that:

$$I_{2h}^h \Psi_{2h}(x) = I_{2h}^h \alpha_{2h} \varphi_{2h}(2\theta, x)$$

$$= \alpha_{2h} \frac{1}{2}(1 + \cos\theta) \varphi_h(\theta, x)$$

$$+ \alpha_{2h} \frac{1}{2}(1 + \cos\bar{\theta}) \varphi(\bar{\theta}, x)$$

and the matrix of the prolongation operator is:

$$\hat{I}_{2h}^h = \frac{1}{2}\begin{pmatrix} 1 + \cos\theta \\ 1 + \cos\bar{\theta} \end{pmatrix}. \tag{16}$$

Two-Grid Operator

The multigrid method [8,15] is a combination between a relaxation method (that reduces very well the high frequency components of the error, but is slowly convergent because of the low frequency components) and the coarse grid correction (which has complementary properties to the smoother).

The matrices from (9), (11), (14), (15) and (16) are used to create the two-grid operator for the multigrid method:

$$\hat{M}_h^{2h} = \hat{S}_h^{v_2} \hat{K}_h^{2h} \hat{S}_h^{v_1} \tag{17}$$

where:

$$\hat{K}_h^{2h} = \hat{I}_h - \hat{I}_{2h}^h \left(\hat{L}_{2h}\right)^{-1} \hat{I}_h^{2h} \hat{L}_h \tag{18}$$

is the matrix of the coarse grid correction operator.

It has been proven [9] that it is sufficient to derive the convergence properties for the two-grid method and the multigrid method will have similar properties. As a consequence, the following factors are defined for the two-grid operator.

Asimptotic convergence factor

$$\rho_{loc}\left(M_h^{2h}\right)$$

$$= \sup\left\{\rho_{loc}\left(\hat{M}_h^{2h}(\theta)\right), \theta \in T^{low} = \left[-\frac{\pi}{2}, \frac{\pi}{2}\right], \theta \notin \Lambda\right\}. \tag{19}$$

Error reduction factor

$$\sigma_{loc}\left(M_h^{2h}\right)$$

$$= \sup\left\{\left\|\left(\hat{M}_h^{2h}(\theta)\right)\right\|, \theta \in T^{low} = \left[-\frac{\pi}{2}, \frac{\pi}{2}\right], \theta \notin \Lambda\right\}.$$

(20)

Here: $\|\cdot\|$ denotes the spectral norm associated with the Euclidian vector norm in $\mathbb{C}^2$, and

$$\Lambda = \left\{\theta \in \left[-\frac{\pi}{2}, \frac{\pi}{2}\right], \tilde{L}_h(\theta) = 0 \text{ or } \tilde{L}_{2h}(\theta) = 0\right\}.$$

(21)

Smoothing factor

$$\mu_{loc}(S_h, v) = \sup\left\{\sqrt[v]{\rho_{loc}\left(\hat{S}_h^{v_2}(\theta)\hat{Q}_h^{2h}\hat{S}_h^{v_1}(\theta)\right)},\right.$$

$$\left.\theta \in T^{low} = \left[-\frac{\pi}{2}, \frac{\pi}{2}\right]\right\}.$$

(22)

where $\rho_{loc}\left(\hat{S}(\theta)\right)$ is the spectral radius of the matrix $\hat{S}(\theta), v = v_1 + v_2$.

Q_h^{2h}, introduced in [9], is an "ideal" coarse grid operator that anihilates the low frequency error components and leaves the high frequency components unchanged:

$$Q_h^{2h}\varphi(\theta, \cdot) = \begin{cases} 0, & \text{if } \theta \in T^{low} = \left[-\frac{\pi}{2}, \frac{\pi}{2}\right] \\ \varphi(\theta, \cdot), & \text{if } \theta = \bar{\theta} \in T^{high} \end{cases}$$

and:

$$\hat{Q}_h^{2h}(\theta) = \begin{pmatrix} 0 & 0 \\ 0 & 1 \end{pmatrix}, \theta \in T^{low}.$$

Since $\rho_{loc}\left(\hat{S}_h^{v_2}(\theta)\hat{Q}_h^{2h}\hat{S}_h^{v_1}(\theta)\right) = \rho_{loc}\left(\hat{Q}_h^{2h}\hat{S}_h^{v}(\theta)\right)$,

$$\mu_{loc}(S_h, v) = \sup\left\{\sqrt[v]{\rho_{loc}\left(\hat{Q}_h^{2h}\hat{S}_h^{v}(\theta)\right)}, \theta \in T^{low}\right\}.$$

(23)

LOCAL FOURIER ANALYSIS RESULTS FOR THE STUDIED PROBLEM

Smoothing Factor

If $\omega = 1$ then the matrix (11) of the smoother after $v = v_1 + v_2$ steps is:

$$\hat{S}_h^v(\theta) = \frac{1}{4^{v-1}}(a-b)^{2(v-1)}\hat{S}_h(\theta)$$

and has the eigenvalues:

$$\lambda_1 = 0, \lambda_2 = \frac{(a-b)^{2(v-1)}}{2^{2v}}\left[(b+1)^2 - (a-1)(b-1)\right],$$ a and b being given in (10),

$$|\lambda_2|^2 = \frac{\left[(q_1-q_2)^2 + 4q_1q_2\cos^2\theta\right]^{2v-1}}{4q_0^{4v}}$$

$$\cdot\frac{\left[q_0^2 + (q_1-q_2)^2 + 4q_1q_2\cos^2\theta + 2q_0(q_1+q_2)\cos\theta\right]}{4q_0^{4v}}$$

For $\theta \in T^{low}$, the eigen value λ_2 attains its maximum absolute value:

$$|\lambda_2| = \left|\frac{q_0 + q_1 + q_2}{2q_0}\left(\frac{q_1+q_2}{q_0}\right)^{2v-1}\right|$$

(24)

for $\theta = 0$. The smoothing factor for the problem (1) is:

$$\mu_{loc}(S_h, v) = \left(\frac{q_1+q_2}{q_0}\right)^2 \sqrt[v]{\frac{q_0+q_1+q_2}{2(q_1+q_2)}}.$$

(25)

Here, q_0, q_1, q_2 are the coefficients given in (5)-(7).

For a = 10^{-4}, ct = 1, d_1 = 1 × 10^{-12}, d_2 = 1 × 10^{-10}, d_3 = 3 × 10^{-10}, v_1 = 1 × 10^{-9}, v_2 = 1 × 10^{-6}, v_3 = 1 × 10^{-6}, the smoothing factors for the Gauss-Seidel relaxation method are presented in **Table 1**.

Table 1: The smoothig factor as a function of $v = v_1 + v_2$ and l.

| | **Finite differences** | | | | |
	$v = 1$	$v = 2$	$v = 3$	$v = 4$	$v = 5$
$l = 3$	0.0058	8.8272×10^{-4}	4.7260×10^{-4}	3.4581×10^{-3}	2.8670×10^{-3}
$l = 4$	0.0215	0.0066	0.0044	0.0037	0.0032
$l = 5$	0.0667	0.0409	0.0348	0.0321	0.0306
$l = 6$	0.1225	0.1504	0.1611	0.1667	0.1701
$l = 7$	0.1250	0.2296	0.3098	0.3598	0.3937
	Finite element method				
	$v = 1$	$v = 2$	$v = 3$	$v = 4$	$v = 5$
$l = 3$	0.3493	0.2801	0.2603	0.2509	0.2454
$l = 4$	0.2809	0.2125	0.1937	0.1849	0.1798
$l = 5$	0.0991	0.0533	0.0434	0.0391	0.0368
$l = 6$	0.1042	0.0955	0.0928	0.0915	0.0907
$l = 7$	0.1250	0.2291	0.3031	0.3486	0.3792

The data from **Table 1** show that:

Ÿ the Gauss-Seidel red-black relaxation method is a very good smoother for this problem as the smoothing factors in the cases presented here are ≤ 0.5 ; Ÿ both the discretization methods lead to good smoothing factors. The finite element method seems slightly more appropriate when the number of grids used in the multigrid method is bigger; Ÿ the number of relaxation steps before and after the coarse grid correction should not be too big as the smoothing factor increases with v.

Asimptotic Convergence Factor and Error Reduction Factor

For the multigrid method having the components described in (9-16), the matrix of the two-grid operator for the problem (1) is:

$$\hat{M}_h^{2h} = \hat{S}_h^{v_2}(\theta) \left[\begin{pmatrix} 1 & 0 \\ 0 & 1 \end{pmatrix} - \frac{1}{4\tilde{L}_{2h}(2\theta)} \right.$$

$$\left. \cdot \begin{pmatrix} (1 + \cos\theta)^2 \, \tilde{L}_h(\theta) & (1 - \cos\theta)^2 \, \tilde{L}_h(\bar{\theta}) \\ (1 - \cos\theta)^2 \, \tilde{L}_h(\theta) & (1 + \cos\theta)^2 \, \tilde{L}_h(\bar{\theta}) \end{pmatrix} \right] \hat{S}_h^{v_1}(\theta)$$

$$(26)$$

For $\theta \in \left[-\frac{\pi}{2}, \frac{\pi}{2} \right]$ the corresponding asimptotic convergence factor and error reduction factor have been computed from the matrix (26) and are given in **Table 2** for different numbers of preand postsmoothing steps.

The data from **Table 2** show that the multigrid method is very rapidly convergent: if at least one smoothing step is performed before and after the coarse grid correction, then the error is reduced by at least a 10^{-3} factor per multigrid cycle.

NUMERICAL RESULTS

The problem (1) has been solved on a domain containing tree layers with different diffusion and convection coefficients ([16,17]).

The error was computed for the exact solution

$$u_{ex}(x,t) = x^2 + t, \max\left(\left|ue(x,t)\right|\right) = 1.44 \times 10^3,$$

$$x \in \Omega = [0, 1620 \text{ nm}], t \in [0, 24 \text{ min}]$$

The time step in the discretization process has been $dt = 60\,s$. **Figure 1**, **Figure 2** and the **Table 3** represent the error after eight multigrid cycles, with two smoothing steps before and two after the coarsegrid correction.

Table 2:. Asimptotic convergence factor and erroon facto $(l = 6)$

Number of smoothig steps	Finite differences		Finite element method	
	$\rho_{loc}(M_s^{2h})$	$\sigma_{loc}(M_s^{2h})$	$\rho_{loc}(M_s^{2h})$	$\sigma_{loc}(M_s^{2h})$
$v_1 = 0, v_2 = 1$	0.1224	0.1731	0.0153	0.1989
$v_1 = 1, v_2 = 0$	0.1224	0.3297	0.0153	0.2420
$v_1 = 1, v_2 = 1$	0.0225	0.0570	9.1275×10^{-4}	0.0037
$v_1 = 2, v_2 = 1$	0.0041	0.0105	6.8022×10^{-5}	2.5397×10^{-4}
$v_1 = 2, v_2 = 2$	7.5569×10^{-4}	0.0019	5.6099×10^{-6}	2.0141×10^{-3}
$v_1 = 3, v_2 = 2$	1.3862×10^{-4}	3.5209×10^{-4}	4.8333×10^{-7}	1.7152×10^{-6}

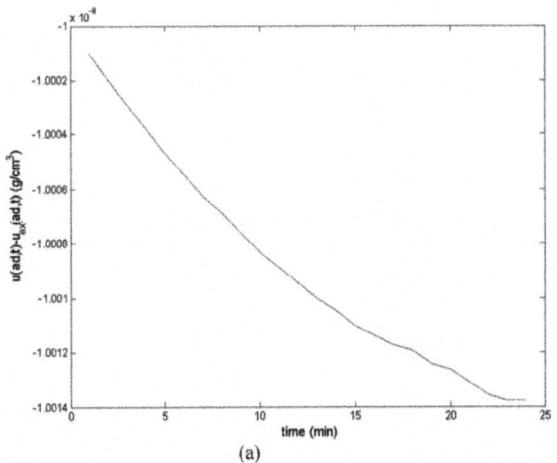

(a)

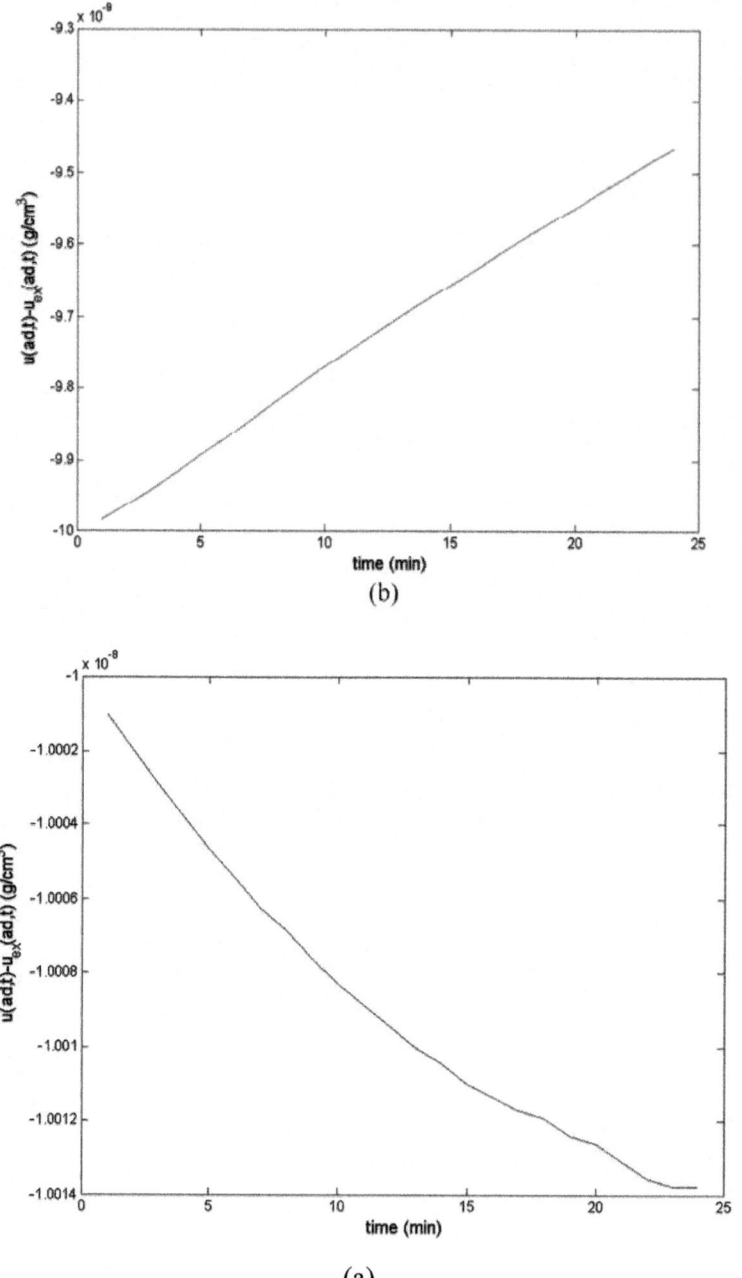

(b)

(a)

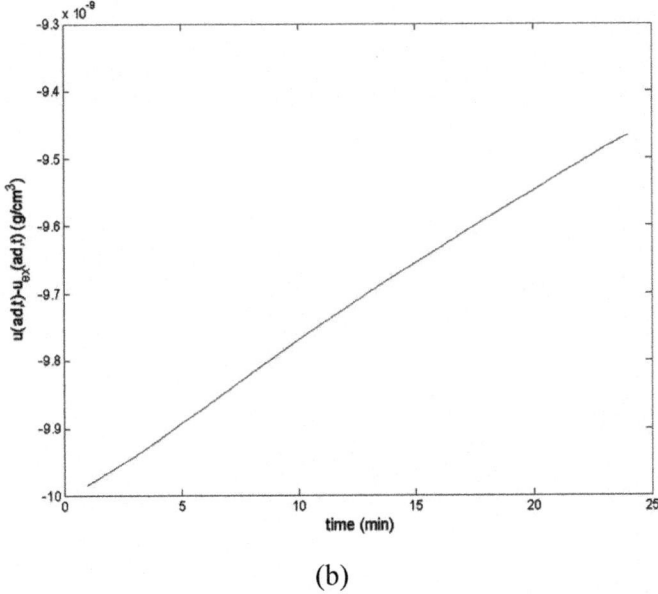

(b)

Figure 1.: The multigrid error at ad = 100 nm in the skin for $v_1 = 1.0 \times 10^{-10}$, $v_2 = 1.0 \times 10^{-7}$, $v_3 = 1.0 \times 10^{-7}$, $d_1 = 1 \times 10^{-12}$, $d_2 = 1 \times 10^{-10}$, $d_3 = 3 \times = 10^4$; $= 0$.

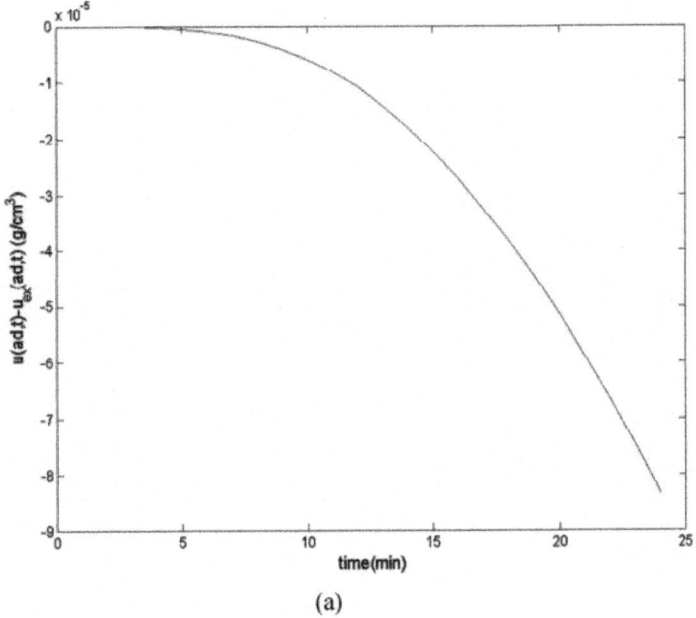

(a)

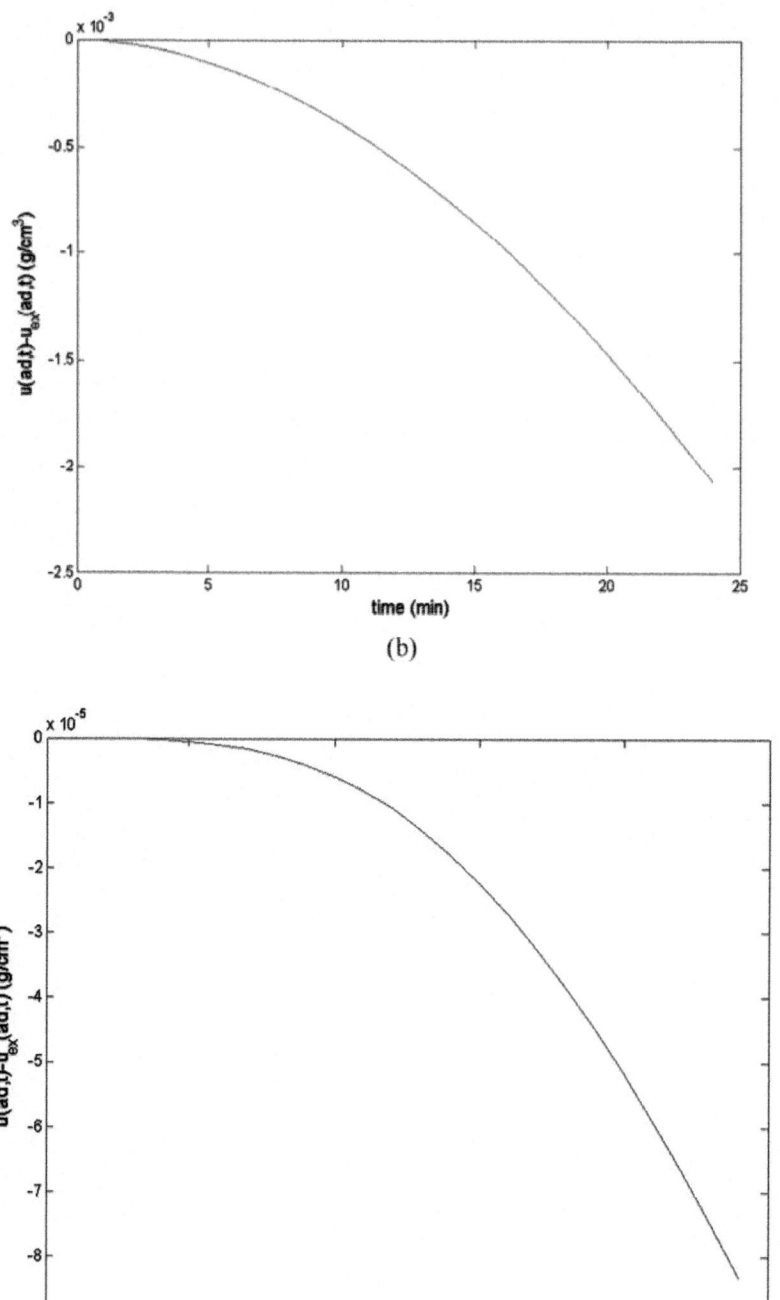

(b)

(a)

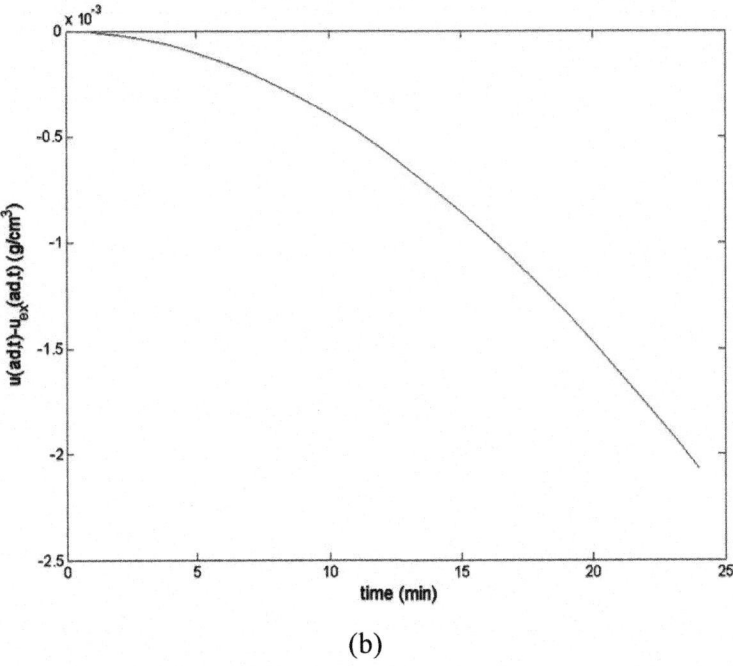

(b)

Figure 2.: The multigrid error at ad = 100 nm in the skin for $v_1 = 1.0 \times 10^{-9}$, $v_2 = 1.0 \times 10^{-6}$, $v_3 = 1.0 \times 10^{-6}$, $d_1 = 1 \times 10^{-12}$, $d_2 = 1 \times 10^{-10}$, $d_3\ 3 \times 10^{10}$; $c = 10^3$; $a = 0$.

	FD	
	$\max_i \lvert u(x_i) - u_{ex}(x_i) \rvert$	$\lVert u - u_{ex} \rVert$
$l = 3$	1.0000×10^{-8}	4.8970×10^{-6}
$l = 4$	1.0057×10^{-8}	3.1812×10^{-8}
$l = 5$	1.0617×10^{-7}	1.6775×10^{-7}
$l = 6$	8.3356×10^{-7}	1.9383×10^{-6}
$l = 7$	5.5076×10^{-6}	1.4308×10^{-5}
	FEM	
	$\max_i \lvert u(x_i) - u_{ex}(x_i) \rvert$	$\lVert u - u_{ex} \rVert$
$l = 3$	9.9847×10^{-9}	4.7554×10^{-8}
$l = 4$	1.6648×10^{-8}	5.9871×10^{-8}
$l = 5$	0.1079	0.2457
$l = 6$	13.3535	30.8845
$l = 7$	104.0864	245.1033

Table 3.: ltigrid error for FD and FEM.

FD		
	$\max_i \left| u(x_i) - u_{ex}(x_i) \right|$	$\| u - u_{ex} \|$
$l = 3$	1.0000×10^{-8}	4.8970×10^{-8}
$l = 4$	1.0057×10^{-8}	3.1812×10^{-8}
$l = 5$	1.0617×10^{-7}	1.6775×10^{-7}
$l = 6$	8.3356×10^{-7}	1.9383×10^{-6}
$l = 7$	5.5076×10^{-6}	1.4308×10^{-5}
FEM		
	$\max_i \left| u(x_i) - u_{ex}(x_i) \right|$	$\| u - u_{ex} \|$
$l = 3$	9.9847×10^{-9}	4.7554×10^{-8}
$l = 4$	1.6648×10^{-8}	5.9871×10^{-8}
$l = 5$	0.1079	0.2457
$l = 6$	13.3535	30.8845
$l = 7$	104.0864	245.1033

Table 3 shows the maximum absolute value of the error and the norm of the error vector corresponding to Figures 1(a) and (b), for the finite differences discretization method (FD) aninite element method (FEM).

CONCLUSION

We have presented a convergence and error analysis for the multigrid method applied to a time dependent diffusion-convection problem that is convection dominated. The mathematical model is applied to the study of the concentration of a solute that is transported by the blood, or the penetration of a substance through the skin layers.

The convergence analysis showed that the discretization process is better realized by the finite element method than the finite differences. Also the red-black Gauss Seidel is a good smoother for the problem presented here, and needs not to be applied more than two or three the constrthe multigridmultigrain method.Themethod. The numerical results in the previous paragraph confirmed the good convergence and error reduction as predicted by the coefficients computed with the local Fourier analysis.

REFERENCES

1. K. W. Morton and D. F. Mayers, "Numerical Solution of Partial Differential Equations, An Introduction," Cambridge University Press, Cambridge, 2005.doi:10.1017/CBO9780511812248

2. E. Becker, G. Carey and J. Oden, "Finite Elements. An Introduction," Prentice-Hall, Englewood Cliffs, 1981.

3. H. C. Elman, D. J. Silvester and A. J. Wathen, "Finite Elements and Fast Iterative Solvers," Oxford University Press, Oxford, 2005.

4. R. P. Fedorenko, "A Relaxation Method for Solving Elliptic Difference Equations," USSR Computational Mathematics and Mathematical Physics, Vol. 1, 1962, pp. 1092- 1096.

5. R. P. Fedorenko, "The Speed of Convergence of One Iterative Process," USSR Computational Mathematics and Mathematical Physics, Vol. 4, 1964, pp. 227-235.

6. A. Brandt, "Multilevel Adaptive Solutions to Boundary Value Problems," Mathematics of Computation, Vol. 31, 1977, pp. 333-390. doi:10.1090/S0025-5718-1977-0431719-X

7. A. Brandt, "Multigrid Techniques: 1984 Guide with Applications to Fluid Dynamics," GMD-Studien Nr. 85, Gesellschaft für Matematik und Datenverarbeitung, St. Augustin, Bonn, 1984.

8. W. Hackbush, "Elliptic Differential Equations," SpringerVerlag, New York, 1992.doi:10.1007/978-3-642-11490-8

9. U. Trottenberg, C. Oosterlee and A. Schuller, "Multigrid," Elsevier Academic Press, London, 2001.

10. P. Weseling, "An Introduction to Multigrid Method," John Wiley & Sons, New York, 1991.

11. M. A. Olshanski and A. Reusken, "On a Robust Multigrid Method for Connection-Diffusion Finite Element Problems." http://www.math.uh.edu/ molshan/ftp/pub/proceed_cd.pdf

12. A. Reusken, "Convergence Analysis of a Multigrid Method for Convection-Diffusion Equations," Numerische Mathematik, Vol. 91, No. 2, 2002, pp. 323-349.doi:10.1007/s002110100312

13. R. Wienands and W. Joppich, "Practical Fourier Analysis for Multigrid Methods," Chapman& Hall/CRC Press, Boca Raton, 2005.

14. R. W. Lewis, P. Nithiarasu and K. N. Seetharamu, "Fundamentals of the Finite Element Method for Heat and Fluid Flow," John Wiley & Sons Ltd, The Atrium, 2004.doi:10.1002/0470014164

15. W. L. Briggs, V. E. Henson and S. McCormick, "A Multigrid Tutorial,"2nd Edition, Siam, Philadelphia, 2000. doi:10.1137/1.9780898719505

16. D. Neumann, "Modeling Transdermal Absorption," Biotechnology: Pharmaceutical Aspects, Springer, New York, 2008.

17. B. Al-Qallaf, D. Bhusan Das, D. Mori and Z. Cui, "Modelling Transdermal Delivery of High Molecular Weight Drugs from Microneedle Systems," Philosophical Transactions of the Royal Society A, Vol. 365, 2007, pp. 2951- 2967. doi:10.1098/rsta.2007.0003

Chapter 11

THE EVOLUTION OF PORE WATER PRESSURE IN A SATURATED SOIL LAYER BETWEEN TWO DRAINING ZONES BY ANALYTICAL AND NUMERICAL METHODS

Abib Tall[1], Cheikh Mbow[2], Daouda Sangaré[3], Mapathé Ndiaye[1], Papa Sanou Faye[1]

[1]Laboratoire de Mécanique et Modélisation, UFR Sciences de l'Ingénieur, Université de Thiès, Thiès, Sénégal

[2]Groupe de Recherches sur les dynamiques des Systèmes et la Mécanique des Fluides, Faculté des Sciences et Techniques, Université Cheikh Anta Diop, Dakar, Sénégal

[3]Laboratoire d'Analyse Numérique et d'Informatique, UFR Sciences Appliquées et Technologie, Université

ABSTRACT

The building of the infrastructure on the compressible and saturated soils presents sometimes major difficulties. The infrastructure undergoes strong settlement that can be due to several phenomena of consolidation of the soils. The latter results from the dissipation of the excess pore pressure and deformation of the solid skeleton. Terzaghi theory led to the equation modeling the dissipation of excess pore pressure. The objective of this study is to establish solutions, by analytical and numerical method, of the equation of the pore water pressure. We considered a compressible saturated soil layer, between two drainage areas and subjected to a uniform load. Separation of variables is used to obtain an analytical solution and the finite element method for the numerical solution. The results obtained by the finite element method have validated those of analytical resolution.

INTRODUCTION

The study of settlements problems of structures built on compressible and saturated soils is generally performed on the basis of theory of the one-dimensional consolidation of Terzaghi [1]. The analysis of the exact solution of the fundamental equation of this theory has aroused many research works among which those of Francesco [2] have combined the solutions of D'Alembert, Fourier and Laplace equations. Work of Ndiaye [3] showed a solution of the equation by the transform of Fourier. Callaud [4] solved the problem with the transform of Laplace. The comparison of the results to those obtained previously had presented offsets.

The objective of this study is to establish analytical and numerical expressions of pore water pressures. For this we will use:

- analytically separate variables method;
- numerically finite element method.

We will consider a compressible saturated soil layer, comprised between two draining areas and subjected to a uniform loading. The resolution of the equation modeling the phenomenon will allow us to predict the evolution of pore water pressure in any point of the layer. For the validation of results obtained in different methods, we make the comparison of curves depending on the space and time.

MATHEMATICAL MODELING OF THE PROBLEM

The consolidation of soils is the physical phenomenon leading to the dissipation of pore water pressure and the deformation of the solid skeleton after application of a load to the surface (seeFigure 1). The study of this phenomenon is very complex especially for the compressible soils because of the low permeability and of the variation of the physical characteristics of the milieu in the course of time.

We will consider a compressible saturated soil layer, comprised between two draining areas and subjected to a uniform loading (see Figure 2). The resolution of the equation modeling the phenomenon will allow us to predict the evolution of pore water pressure in any point of the layer.

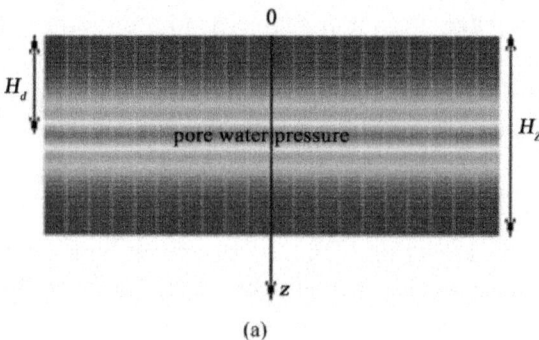

(a)

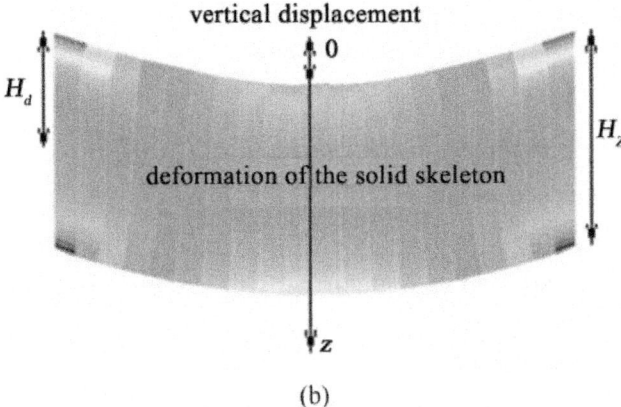

(b)

Figure 1: Principle phenomenon of consolidation of saturated soils. (a) Dissipation of pore water pressure; (b) Deformation of solid skeleton.

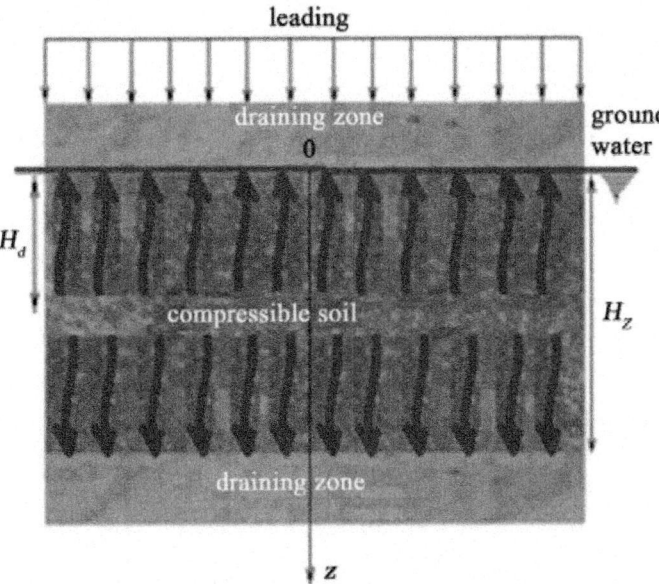

Figure 2: Compressible and saturated layer soil comprised between two draining areas (modified from [1]).

To study this problem then, they make use of simplifying assumptions. On the basis of these working hypothesis then the phenomenon studied is governed by

$$c_v \frac{\partial^2 u(z,t)}{\partial z^2} = \frac{\partial u(z,t)}{\partial t}$$

(1)

With

c_v: vertical coefficient of consolidation;

z: vertical space variable;

t: time.

For the unicity of the solution we will associate the Equation (1) the following conditions:

- initial conditions

$$t = t_0 \quad u(z,t_0) = u_0(z)$$

(2)

- boundary conditions

$$z = 0 \quad u(0,t) = 0$$

(3)

$$z = H_z \quad u(H_z,t) = 0$$

(4)

To generalize this study and facilitate the numerical resolution, we will non-dimensional our equations through the introduction of reference variables:

- vertical non-dimensional thickness

$$Z = \frac{z}{H_z}$$

(5)

- vertical non-dimensional time

$$T_z = \frac{t}{t_r}$$

(6)

The thickness of the compressible soil layer H_z is equal to twice the distance from drainage H_d

$$H_z = 2H_d$$

(7)

The previous works performed by Schiffman [5], Legrand et al. [6] and Skempton et al. [7] showed that the factor of vertical time is given by the expression:

$$T_v = \frac{c_v \cdot t}{\left(2H_d\right)^2}$$

(8)

Equation (1) becomes a non-dimensional relation in the following form:

$$\frac{\partial^2 u(Z,T_z)}{\partial Z^2} = \frac{c_v \cdot t_r}{H_z^2} \frac{\partial u(Z,T_z)}{\partial T_z}$$

(9)

While posing

$$T_{oz} = \frac{c_v \cdot t_r}{H_z^2}$$

(10)

The relation (9) becomes

$$T_{oz} \frac{\partial^2 u(Z,T_z)}{\partial Z^2} = \frac{\partial u(Z,T_z)}{\partial T_z}$$

(11)

The non-dimensional initial and boundary conditions associated are:

- non-dimensional initial conditions

$$u(Z,T_z) = u_0(Z)$$

(12)

- non-dimensional boundary conditions

$$u(0,T_z) = 0$$

(13)

and

$$u(1,T_z) = 0$$

(14)

METHODS OF RESOLUTION

For each method of resolution, we will perform a non-dimensional transformation of the studied partial derivative equation. This will give us a non-dimensional time proportional to the vertical time factor. We will use analytically the separate variables method and numerically the finite element method.

Separated Variables Method (SVM)

The resolution of the Equation (11) by the method of the separated variables is used in several works; mention may be made the results of Braja [8] and Magnan [9]. The Equation (11) and its boundary conditions are solved by using the separated variables method. It allows obtaining an analytical solution in the form of a product of functions

$$u(Z,T_z) = u(Z) \cdot \eta(T_z)$$

(15)

While replacing (15) in (11), they obtains

$$T_{oz} \frac{1}{u(Z)} \frac{\partial^2 u(Z)}{\partial Z^2} = \frac{1}{\eta(T_z)} \frac{\partial \eta(T_z)}{\partial T_z}$$

(16)

From where

$$\frac{1}{u(Z)} \frac{\partial^2 u(Z)}{\partial Z^2} = -C^2$$

(17)

and

$$\frac{1}{\eta(T_z)} \frac{\partial \eta(T_z)}{\partial T_z} = -C^2 T_{oz}$$

(18)

with C a constant. After having applied the initial and boundary conditions posed into 2, they obtain the expression of the pore water pressure in the form:

$$u(Z,T_z) = 4 \sum_{n=0}^{+\infty} \frac{u_0(Z)}{N_n} \sin(N_n Z) \exp\left(-(N_n)^2 T_{oz} T_z\right)$$

(19)

with

$$N_n = (2n+1)\pi$$

(20)

While posing

$$T_{zv} = T_{oz} T_z$$

(21)

From where the expression of the pore water pressure

$$u(Z,T_z) = 4\sum_{n=0}^{+\infty} \frac{u_0(Z)}{N_n} \sin\left(N_n \frac{z}{H_z}\right) \exp\left(-(N_n)^2 T_{zv}\right)$$

(22)

Finite Element Method (FEM)

This method of resolution is used in many studies; they can quote work of Merrien [10], Goncalvès [11] and Dhatt et al. [12]. The finite elements used to obtain an approximate value of the solution of the Equation (11). For the resolution, we considered a linear reference element of Lagrange type. The strong variational formulation gives the following relation.

$$T_{oz} \cdot \eta(T_z) \cdot \left(\int_0^1 \nabla u(Z) \cdot \nabla v(Z) \cdot dZ - \left[\nabla u(Z) \cdot v(Z)\right]_0^1 \right) + \frac{\partial \eta(T_z)}{\partial T_z} \cdot \int_0^1 u(Z) \cdot v(Z) \cdot dZ = 0$$

(23)

The following relation being null

$$\left[\nabla u(Z) \cdot v(Z)\right]_0^1 = 0$$

(24)

The weak variational formulation gives the following relation

$$T_{oz} \cdot \eta(T_z) \cdot \int_0^1 \nabla u(Z) \cdot \nabla v(Z) \cdot dZ + \frac{\partial \eta(T_z)}{\partial T_z} \cdot \int_0^1 u(Z) \cdot v(Z) \cdot dZ = 0$$

(25)

The approximate value is given by the expression

$$\bar{u}(\xi) = \sum_{j=1}^N u_j N_j(\xi_i)$$

(26)

With $N_j(\xi_i)$: shape function. The combination of the variational formulation and the shape function gives the matrices of following elementary mass and rigidity:

$$M_{ze} = \int_{-1}^1 [N(\xi)] \cdot [N(\xi)]^{\mathrm{T}} \frac{\partial Z}{\partial \xi} d\xi$$

(27)

from where the elementary mass matrix

$$M_{ze} = \frac{Z_j - Z_i}{6} \begin{pmatrix} 2 & 1 \\ 1 & 2 \end{pmatrix}$$

(28)

and

$$R_{ze} = \int_{-1}^{1} [N'(\xi)] \cdot [N'(\xi)]^{T} \frac{\partial Z}{\partial \xi} d\xi$$

(29)

from where the elementary rigidity matrix

$$R_{ze} = \frac{1}{Z_j - Z_i} \begin{pmatrix} 1 & -1 \\ -1 & 1 \end{pmatrix}$$

(30)

The assembly of the elementary matrix makes it possible to determine the solution approached to the whole domain.

$$M_{ze} = \frac{Z_j - Z_i}{6} \begin{pmatrix} 2 & 1 & 0 & 0 & 0 \\ 1 & 4 & \ddots & 0 & 0 \\ 0 & \ddots & \ddots & \ddots & 0 \\ 0 & 0 & \ddots & 4 & 1 \\ 0 & 0 & 0 & 1 & 2 \end{pmatrix}$$

(31)

and

$$R_{ze} = \frac{1}{Z_j - Z_i} \begin{pmatrix} 1 & -1 & 0 & 0 & 0 \\ -1 & 2 & \ddots & 0 & 0 \\ 0 & \ddots & \ddots & \ddots & 0 \\ 0 & 0 & \ddots & 2 & -1 \\ 0 & 0 & 0 & -1 & 1 \end{pmatrix}$$

(32)

The solution of the relation (11) is given in the form of an ordinary differential equation

$$M_{zg} \cdot \frac{\partial \eta(T_z)}{\partial T_z} + T_{oz} R_{zg} \cdot \eta(T_z) = 0$$

(33)

With:

M_{zg}: global matrix of mass;

R_{zg}: global matrix of rigidity.

NUMERICAL Simulation and Analysis

The numerical simulation is based on the ratio of pore pressure and that of its initial value with step of regular grids. The evolution of the pore water pressure obtained by the analytical method (SVM) and numerical (FEM) are represented in the Figure 3(a) and Figure 3(b). The graphs of the Figure 3 shows the isochrones obtained of the Figure 4.

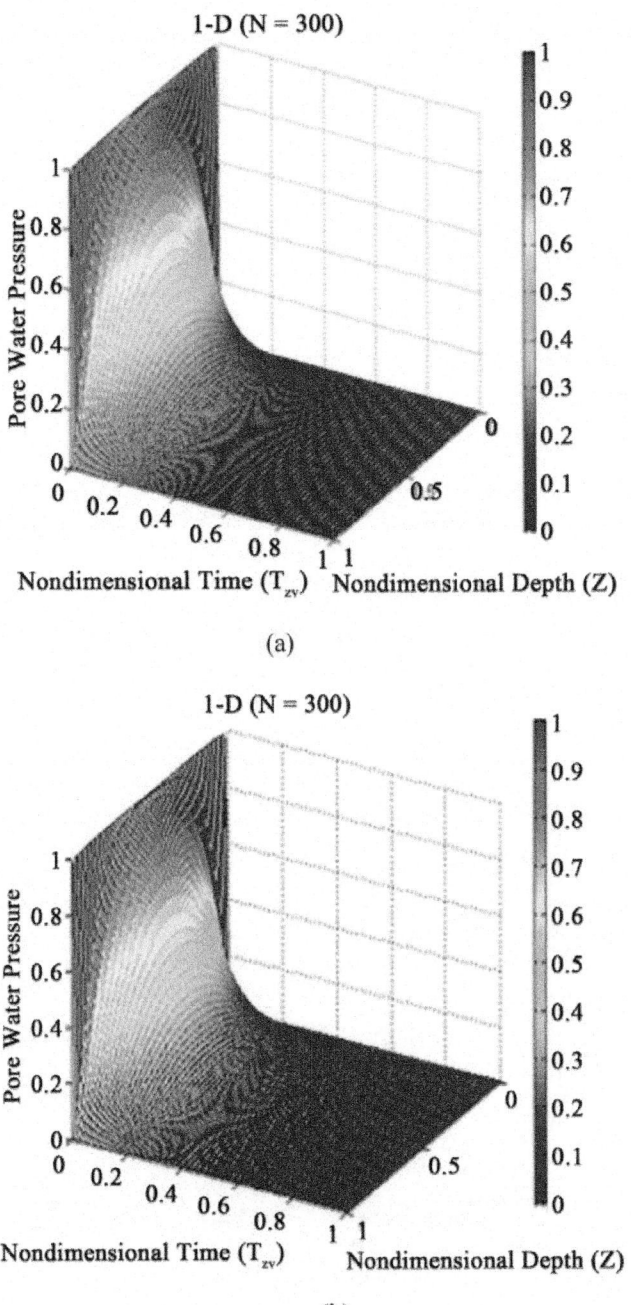

Figure 3: Evolution of ratio pore water pressure according to the non-dimensional variables. (a) Separated variables method; (b) Finite element method.

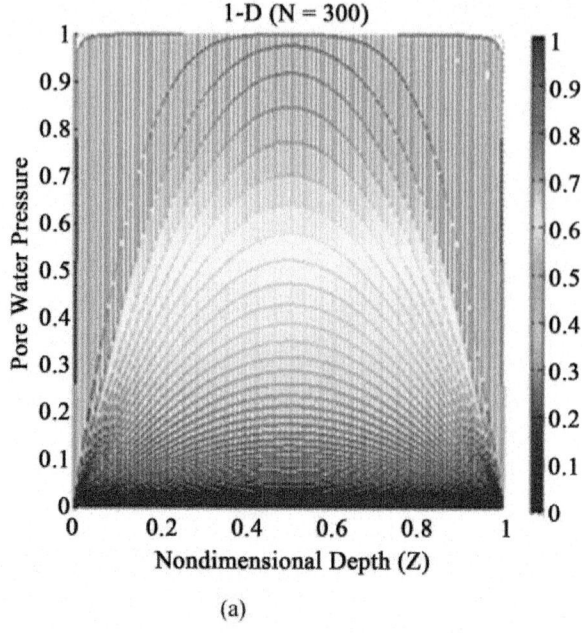

(a)

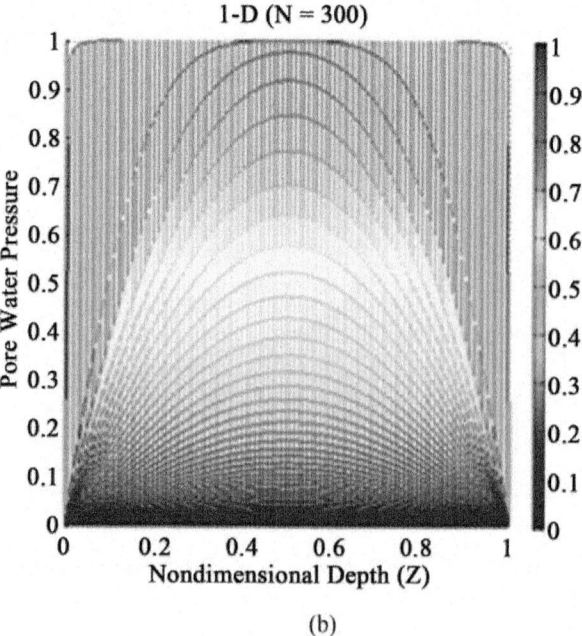

(b)

Figure 4: Isochrones of ratio pore water pressure. (a) Separated variables method; (b) Finite element method.

The evolution of pore pressure depending of non-dimensional time is represented in the Figure 5. We found that the graphs of the Figure 4 and the Figure 5 have the same allure. It appears from this observation that the evolution of pore water pressure as a function of reduced dimensions, obtained by separated variables method, is somewhat similar to those of the finite elements. To assess the reliability of the solutions analytical and numerical we will perform a comparison.

Comparison of the Analytical and Numerical Solutions The isochrones of pore water pressure obtained by the method of the separated variables and the finite element method are represented in the Figure 6 with $T_{zv} = 0$, $T_{zv} = 0.01$ and $T_{zv} = 0.1$. According to the graphs of the Figure 6, we notice that the evolutions of pore water pressure almost superposed. We can note that the solutions obtained by SVM and FEM are similar for each selected time factor. To consider error made between the two solutions exact and approached pore pressure we will carry out a comparison by linear regression (Figure 7). We note that for each graph of the Figure 7, there is a linear relation between the values of analytical and numerical pore pressures. For each graph of the Figure 7, we obtained a linear regression line near to y = x and a coefficient of regression appreciably equal to $R^2 = 1$. So, we can note an almost superposition of analytical and numerical isochrones

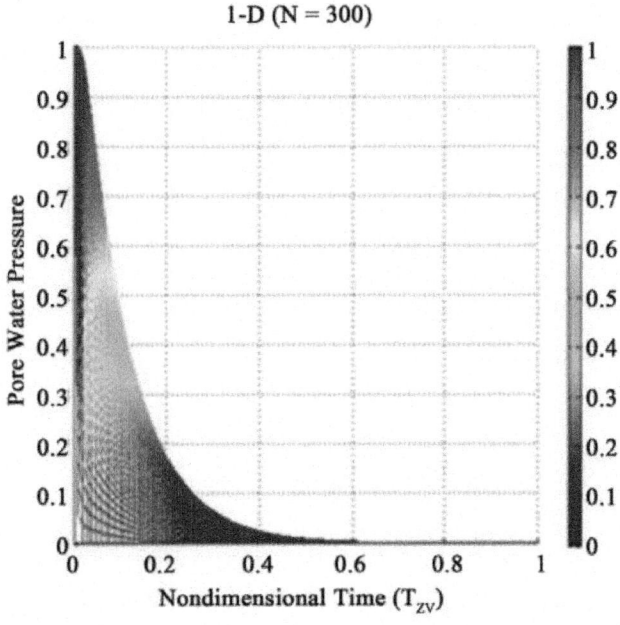

(a)

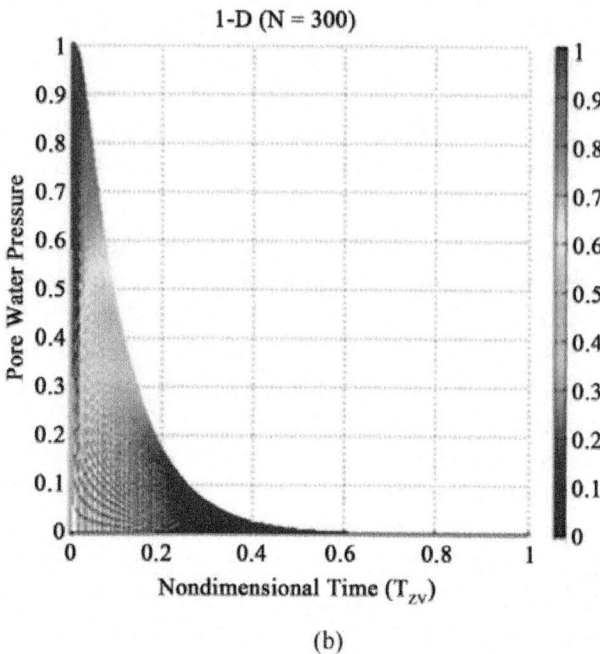

(b)

Figure 5: Evolution of ratio pore water pressure depending of non-dimensional time factor. (a) Separated variables method; (b) Finite element method.

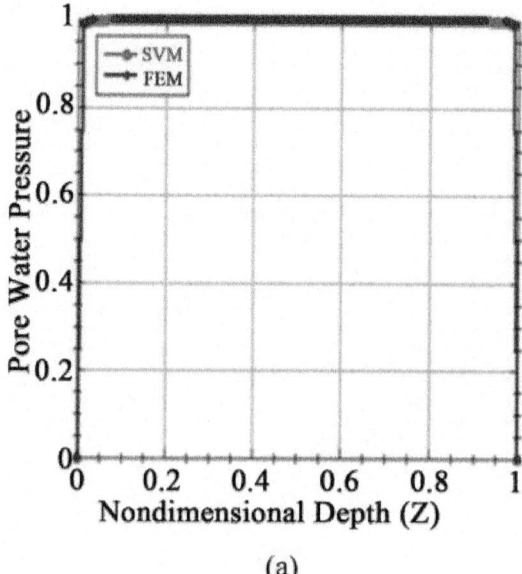

(a)

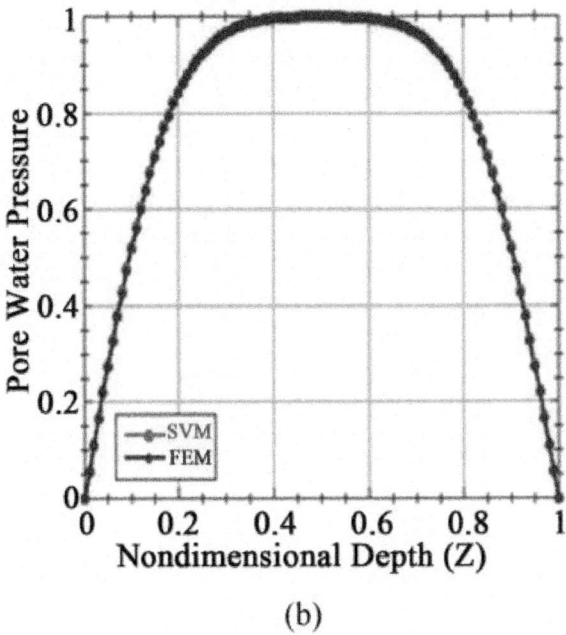

(b)

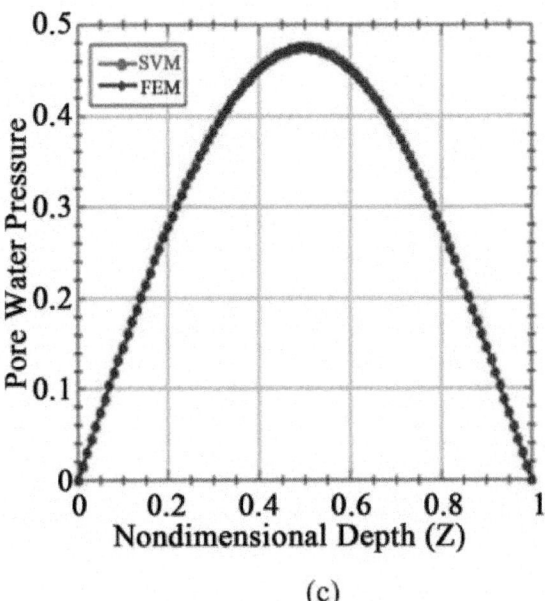

(c)

Figure 6: Comparison of the isochrones of pore water pressure to (a) $T_{zv} = 0$; (b) $T_{zv} = 0.01$ and (c) $T_{zv} = 0.1$.

The study of the value of pore water pressure in a compressible soil layer at each point was given. The curves obtained by of separated variables method and of finite element method are represented in the Figure 8 to Z = 0.25, Z = 0.5 and Z = 0.75. It can note well that the evolutions of pore water pressure as a function to non-dimensional time almost superposed according to the graphs of the Figure 8. For the evaluation of the error made between the solutions exact and approached pore pressure as a function of time, we conducted a comparison of the values obtained in the graphs of Figure 9.

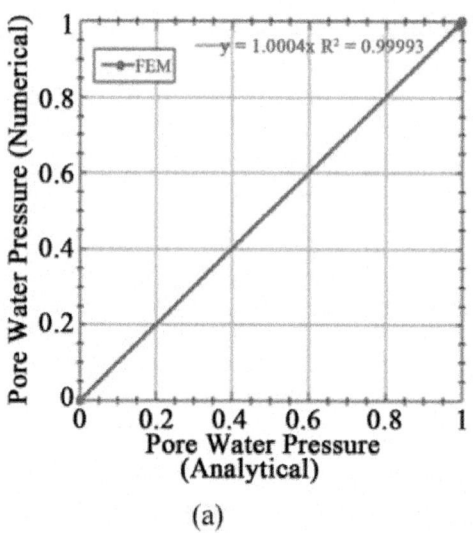

(a)

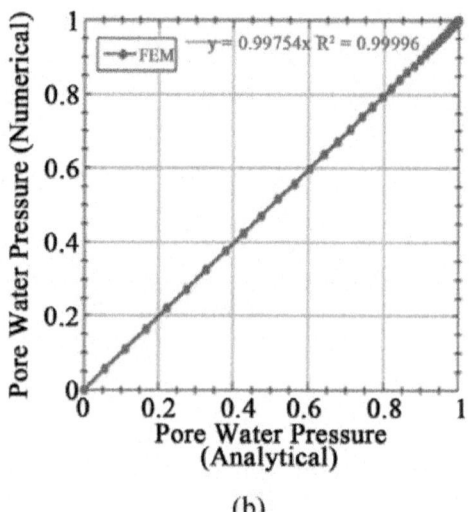

(b)

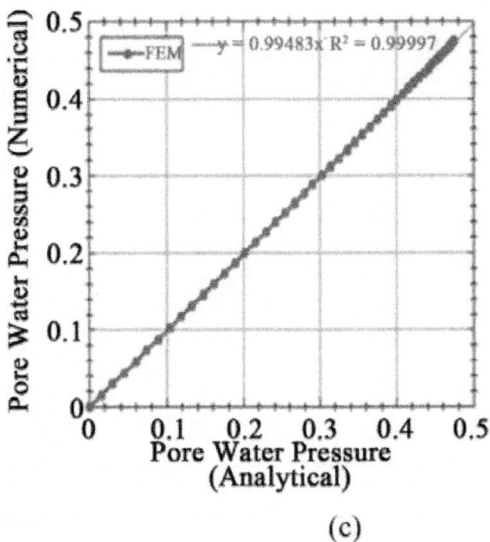

(c)

Figure 7: Evolution of the values of numerical pore water pressure as a function to those analytical to (a) $T_{zv} = 0$; (b) $T_{zv} = 0.01$ and (c) $T_{zv} = 0.1$.

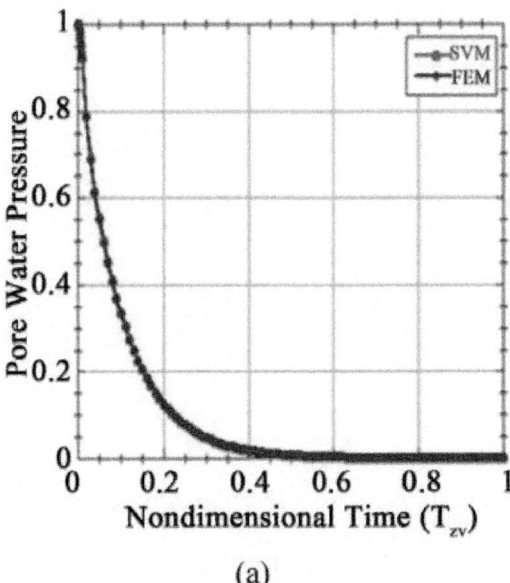

(a)

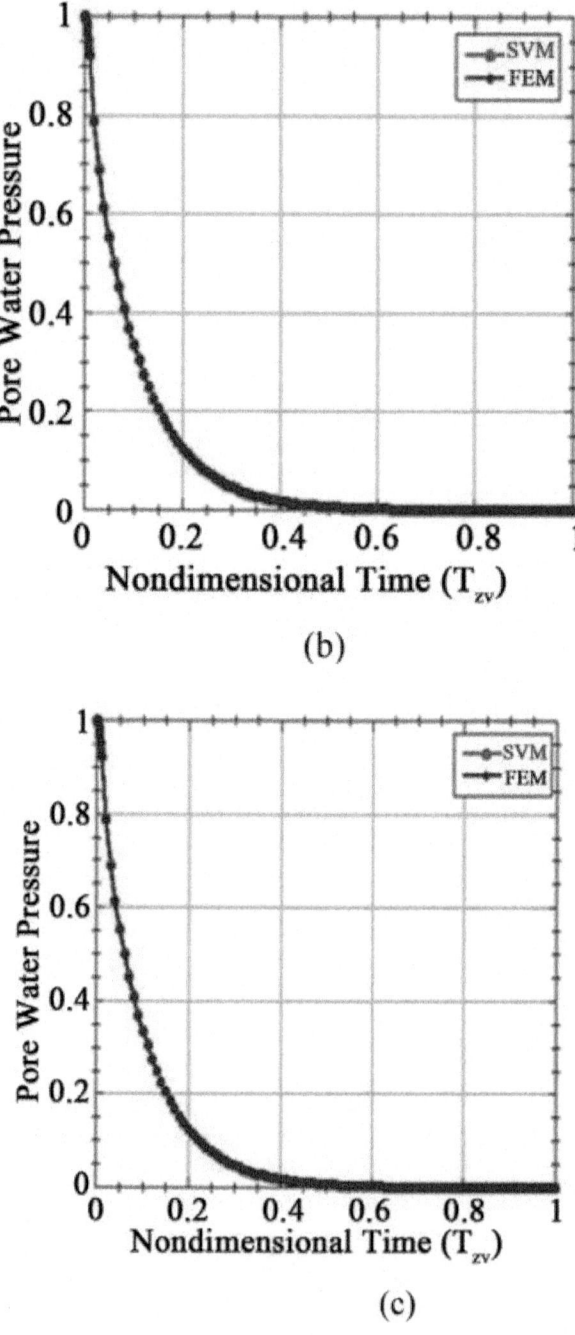

(b)

(c)

Figure 8: Comparison of pore water pressure as a function of non-dimensional time to (a) Z = 0.25; (b) Z = 0.5 and (c) Z = 0.75.

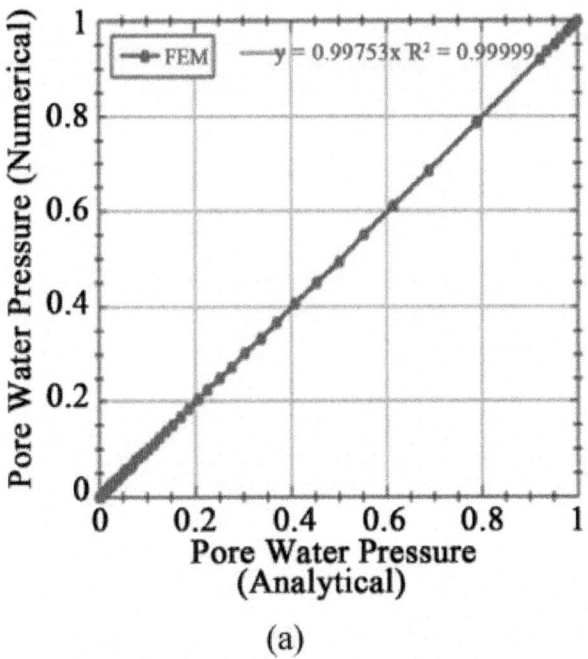

(a)

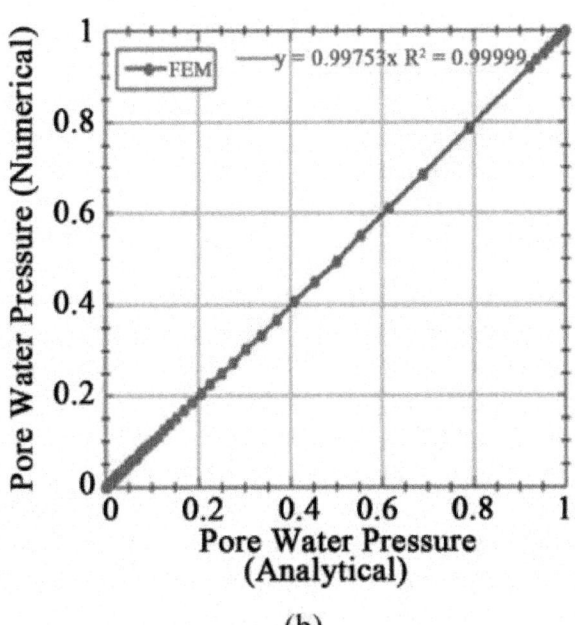

(b)

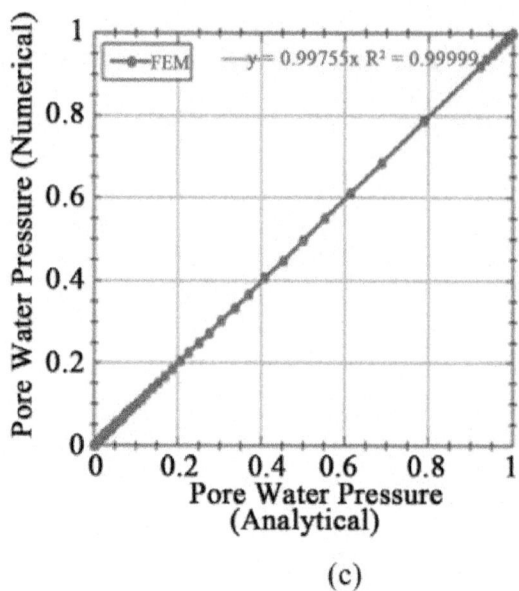

(c)

Figure 9: Evolution of the values of numerical pore water pressure according to those analytical for (a) Z = 0.25; (b) Z = 0.5 and (c) Z = 0.75.

We note that for each graph of the Figure 9, there is a linear relation between the values of analytical and numerical pore water pressures. For each graph of Figure 9, we obtained a linear regression line near to y = x and a coefficient of regression almost equal to $R^2 = 1$. So, we can note an almost superposition of the analytical and numerical of curves.

CONCLUSIONS

This research has been enabled to study the evolution of pore water pressure in a compressible and saturated soil layer, between two draining areas subjecting a uniform loading on the surface. The non-dimensional transformation of time and thickness enabled us to understand and solve the problem. The examination of the analytical solution is obtained by the separate variables method validated by the finite element method; let's say that the results are satisfactory for the resolution of the problems of primary consolidation. A comparative study by linear regression shows that the error is substantially equal to zero with a coefficient of regression close to 1. Finite element method approaches solutions of well separated variables method with respect to the equation from the primary consolidation. We can conclude that, the solutions obtained can be used for the study of pore water pressure in compressible and saturated soil subjecting a uniform load.

ACKNOWLEDGEMENTS

I thank the authors for this work for their contributions and the Group of research of the laboratory of Mechanics and Modeling of the UFR of the engineering of the University of Thies.

REFERENCES

1. Terzaghi, K. (1943) Theoretical Soil Mechanics. Wiley, New York. http://dx.doi.org/10.1002/9780470172766

2. Di Francesco, R. (2011) Exact Solution to Terzaghi's Consolidation Equation. Wizard Technology, Teramo, Italy.

3. Ndiaye, B. (1996) Application de la transformée de Fourier à la résolution de problèmes de couplage hydromécanique. D.E.A de l'Ecole Nationale des Mines de Nancy.

4. Tchouani Nana, J.M. and Callaud, M. (2004) Cours de mécanique des sols. Tome 1, 120.

5. Schiffman, R.L. (1958) Consolidation of Soil under Time Dependent Loading and Varying Permeability. Proceedings of the Thirty-Seventh Annual Meeting of the Highway Research Board, Washington DC, 6-10 January 1958, 584-617.

6. Legrand, J. and Schlosser, F. (1973) Embankment on Compressible Soil. Society of Civil Engineers, 98, 285-312.

7. Skempton, A.W. and Bjerrum, L. (1957) A Contribution to the Settlement Analysis of Foundations on Clay. Géotechnique, 7, 168-178. http://dx.doi.org/10.1680/geot.1957.7.4.168

8. Braja, M.D. (2008) Advanced Soils Mecanics. 3rd Edition, Taylor & Francis, New York.

9. Magnan, J.P. (2000) Déformabilité des sols. Tassement. Consolidation, Technique de l'Ingénieur, C 214.

10. Merriem, J.L. (2007) Numerical Analysis with Matlab. INSA, Rennes, 200-204.

11. Goncalvès, E. (2005) Résolution Numérique. Discrétisation des EDP et EDO, Institut National Polytechnique de Grenoble, Grenoble.

12. Dhatt, G. and Touzot, G. (1981) A Presentation of the Finite Element Method. International Journal for Numerical Methods in Geomechanics Analytical, 5, 1-14

Chapter 12

DESIGN OF OVERALL SLOPE ANGLE AND ANALYSIS OF ROCK SLOPE STABILITY OF CHADORMALU MINE USING EMPIRICAL AND NUMERICAL METHODS

Mahdi Rasouli Maleki[1], Mohammad Mahyar[2], Kambiz Meshkabadi[3]

[1]Engineering Geology & Rock Mechanic Department, Tunnel Consulting Engineers, Tehran, Iran

[2]Mining Engineering, Tunnel Consulting Engineers, Tehran, Iran

[3]Lecturer of Civil Engineering Department, Islamic Azad University of Ahar, Iran

ABSTRACT

In engineering projects associated with rock mechanic science like open pit mines, assessment and slope stability of mine walls is one of the important performance in generate of these structures. Estimating and knowledge of stable slope angle is one of main parts that should be occurring to special attention in open pit mines studies phase. Considering the importance of economic costs in mining issues, the need for appropriate design slope angle that can cause an adverse minimize project costs and throws the other hand, the stability conditions in the safe walls of the mine life will provide essential and seems obvious. Therefore, in this study to determine the optimal slope angle of overall and bench of west wall of the Chadormalu ore iron mine, has been trying, first, done field studies on the discontinuity of western wall, engineering classification and geomechanical properties of rock masses of wall, then assess the amount of optimal slope angle using empirical method. Finally, in order to ensure stability and accuracy of the wall slope angle based on the obtained (empirical method) tries to analysis is amount of Factor of Safety (FOS), displacements and mean stress condition atwalls calculated from drilling use Phase2D powerful software.

INTRODUCTION

The purpose of this study is to determine the bench slope angle and overall slope of the west wall in Chadormalu mine in points susceptible to rupture.

To do so, the survey tries to; first, detect sensitive points by current empirical methods. Then it determines the bench and overall angle of slope.

In order to be sure about the results validity obtained by the empirical methods, the study attempts to analyze stability and determine the slope safety factor and the wall displacements using finite element method and powerful Phase2D software.

POSITION AND GEOLOGY OF THE WEST WALL OF THE MINE

Chadormalu iron ore mine is located in central Iran and in northern slope of Chah-Mohammad grey mountains in southern margin of Saghand salt marsh about 180 km from north-east of Yazd and 300 km from south of Tabas desert. According to the geology studies performed in this region, it was cleared that Chadormalu fault between the plain and high lands is the major factor of ore creation and mineralization in the region formed in two forms of northern and southern anomaly. Also, petrography studies on the mine rocks shows that major rocks in Chadormalu mine area are Metasomatite, Albitite, Diorite, Magnetite and Hematite[4,5]. It should be mentioned that, performance of different faults in this area makes the mine rocks tobe severely tectonizedand provide suitable conditions fordifferent ruptures of the wall.

The western wall of this mine is made up of igneous rocks and various metamorphic rocks such as Diorite, Albitite and Metasomatite[4]. Generally speaking, **Figure 2** shows the geological profile perpendicular in the western wall of the mine together with its lithological combination.

Figure 1: Location of area study on the Iran map.

Profile on the midcourse of block 11 & Block 12

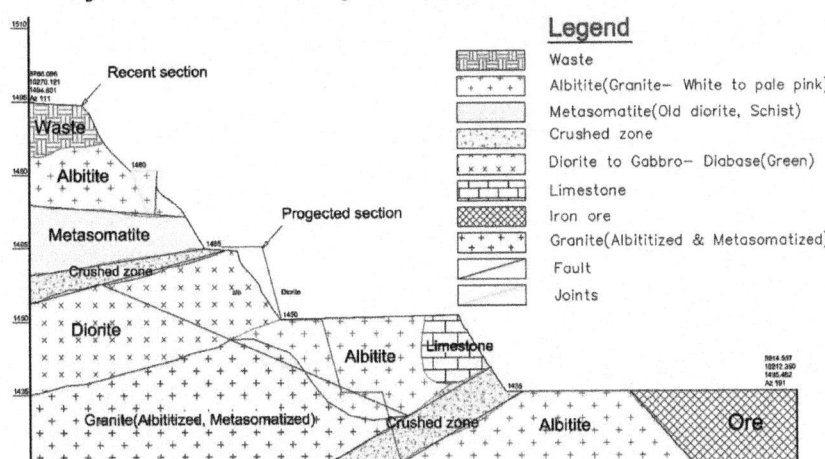

Figure 2: The geological profile perpendicular on the west wall of Chador-malu iron ore mine.

In this study, in order to save time and costs, three blocks (No. B-10, B-11, and B-13) in instability intensity wear detected as the conclusion of the geological surveys on the mine western wall, than engineering surveys and joint studies ware done on each of these blocks [5].

CLASSIFICATION OF ROCK MASS AND DETERMINATION OF ENGINEERING PARAMETERS

As the main purpose of engineering projects is to use classification systems to determine geomechanical characteristics of rocks by simple methods, this study tries to do joint studies on present discontinuities; and then it classifies the rocks enclosed in each of the blocks using Rock Mass Rating (RMR), Geological Strength Index (GSI) and Slope Rock Mass Rating (SRMR) classification systems [3]. **Table 1** indicates the results of the classification of the rocks of the western wall in Chadormalu iron ore mine.

Results obtained from this table indicate that the quality of the rocks in the west wall area in Chadormalu mine is poor due to breakings and development of lots of joints and fractures.

As in engineering works especially in analysis of rock slopes, the purpose is to classify rocks to estimate and measure their engineering and geomechanical features correctly, this study uses results of rocks classification and Roclab software [6] for each of the rock pieces enclosed in block B-11 to detect those parameters that have been introduced by empirical methods established by researches throughout the world. **Table 2** shows the most important calculated engineering parameters which are used in Phase2D software [7].

DETERMINATION OF BENCH SLOPE AND OVERALL SLOPE

According to definition of slope geometry, it is said that height, width, and angle of bench slope are the most significant geometrical parameters of slopes and steep surfaces where any alternation each of these features can put a direct effect on the slope stability. On the basis of these words, therefore, one can admit that optimum determination of these geometrical features of a slope in preventing rupture is one of most important parts of rock and soil slopes analysis, so that importance of this issue in open pit mine activities and road cuttings are observable and underst-andable [2].

This study tries to apply not only ranking system, but also other empirical methods in determina-tion of bench slope and overall steep of the slope in order to promote the obtained results safety factor. Therefore, value of the rocks of each block, one can determine slope steep angle by the empirical methods obtained from Rock Mass Rating (RMR), Mine Rock Mass Rating (MRMR) and Slope Rock Mass Rating (SRMR) values. Results on overall angle and safe bench slope angle for each block are shown in Tables 3 and 4 respectively. Also, **Table 5** indicates features and final geometry of the west wall in Chadormalu mine.

Table 1: Results of the classification of the west wall Chadormalu mine according to various classification systems

System	NO. Block	Metasomatite	Albitite	Diorite	Fault	Crushed zone	Average
	B - 10	28	-	-	25	23	25
RMR	B - 11	26	26	-	-	21	24
	B - 12	39	39	38	-	-	39
	B - 10	23	-	-	20	18	20
GSI	B - 11	21	21	-	-	16	19
	B - 12	34	34	33	-	-	34
	B - 10	52	-	-	45	44	47
SRMR	B - 11	52	49	-	-	44	48
	B - 12	65	62	60	-	-	63

Table 2: The most important engineering parameters of each rock groups

Material	Albitite	Metasomatite	Granite	Diorite	Crushed zone
Unit weight (MN/m³)	0.024	0.028	0.024	0.028	0.026
Compressive Strength of Rock mass (MPa)	2.29	2.52	3.24	3.31	1.83
Young's modulus (MPa)	1632	1691	3901.5	3270	1268
Poisson's ratio	0.26	0.23	0.23	0.24	0.26
Tensile strength (MPa)	0	0	0	0	0
Peak friction angle (degrees)	27.1	26	56.1	33.5	22.2
Peak cohesion (MPa)	0.67	0.68	0.21	1.01	0.54
Dilation Angle (degrees)	0	0	0	0	0
Residual Friction Angle (degrees)	27.1	26	56.1	33.5	22.2
Residual Cohesion (MPa)	0.67	0.68	0.21	1.01	0.54

STABILITY ANALYSES OF THE MINE WEST SLOPE

Introduction

Today, there are several methods for slope stability analysis, each has its own advantages and disadvantages. Numerical analyses methods are the most common ones that are used for rock and soil slope analyses. One of the software's that can analyze rock slope stability in a numerical way is powerful software called Phase2D. This software was used in the present study for analyses of stability in the west wall of Chadormalu mine.

Hypotheses of Analyses

In this research, it is supposed the all analyses have been done in conditions prior to excavation of berm 1435 and for both static and dynamic states whit 0.31 g earthquake acceleration.

Analyses and Results of the Slope Stability

Modeling and bordering the concerned slope in finite element Phase2D software and running the program, result of stability in the west slope of the mine were examined. The results show that safety factor of the slope designed under static and dynamic conditions will be 3.39 m and 2.26 m, respectively (**Table 6**). Also, displacement due to berms excavation is 1.5 cm for static state and 1.6 cm for dynamic one. Figures 3 and 4 represents the outcome model of Phase2D software (Factor of safety, total displacement and mean stress status) in condition prior to excavation of bench 1435 and **Figure 5** shows the shear-strain changes for both static and dynamic states, respectively.

Table 3: Safe overall slope angle obtained by Bieniawski (1989)[1] method

NO. Block	Metasomatite	Albitite	Diorite	Fault	Crushed zone	Average
B - 10	45.6	-	-	41.7	38.7	42.0
B - 11	42.6	42.6	-	-	35.0	40.1
B - 12	57.4	57.4	56.5	-	-	57.1

Table 4: Safe bench slope angle obtained by Slope Rock Mass Rating (SRMR)

NO. Block	Metasomatite	Albitite	Diorite	Fault	Crushed zone	Average
B - 10	69	-	-	66	65	66
B - 11	69	67	-	-	65	67
B - 12	74	73	72	-	-	73

Conclusions

Results obtained from this study confirm that; in dynamic condition, to obtain a safety factor upper than 2.2, the bench slope angle and its overall slope angle should be 70 degree and 44 degree. Also, according to the features of the rock masses in this area, numerical analysis results indicate that; under such a condition, displacement due to bench excavation will be led than 1.5 cm.

Table 5: Features and final geometry obtained for bench for the west wall in Chadormalu mine

Slope Parameters	Value
High (m)	15
Width of bream (m)	8.5
Bench Angle (degree)	70
Inter-ramp Angle (degree)	47
Overall Angle (degree)	44

Table 6: Slope safety factor and displacement due to slopes excavation

Parameters	Static	Dynamic
Safety of Factor (SRF)	3.39	2.26
Horizontal displacement (m)	0.015	0.016

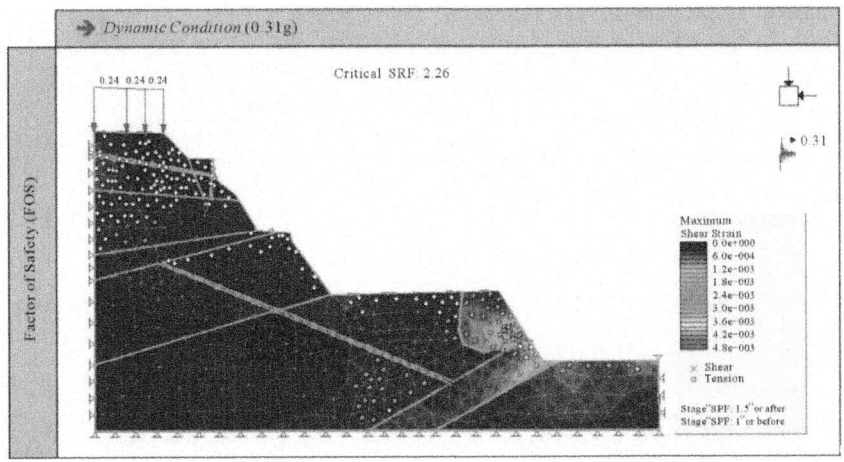

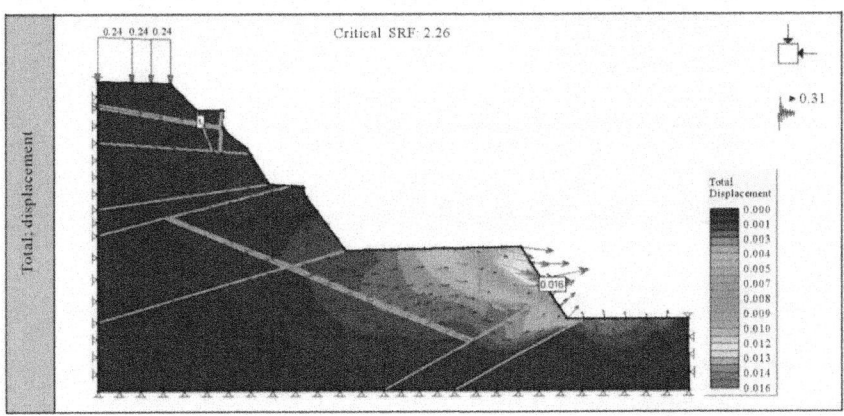

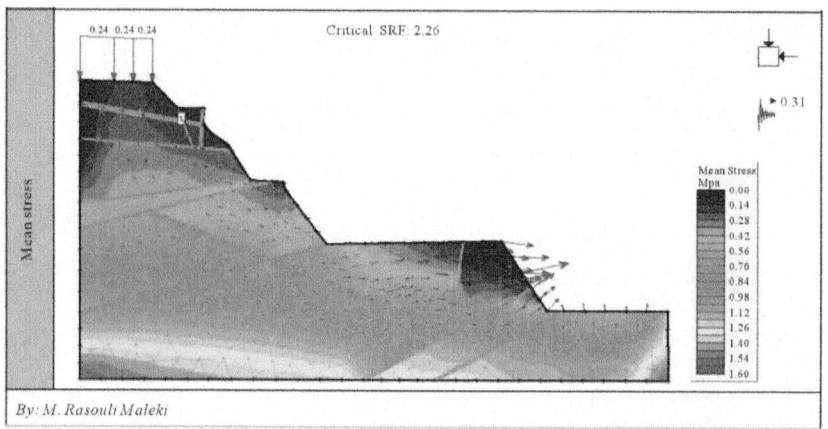

Figure 3: Output model of Phase2D software, factor of safety (FOS), total displacements and mean stress for static state.

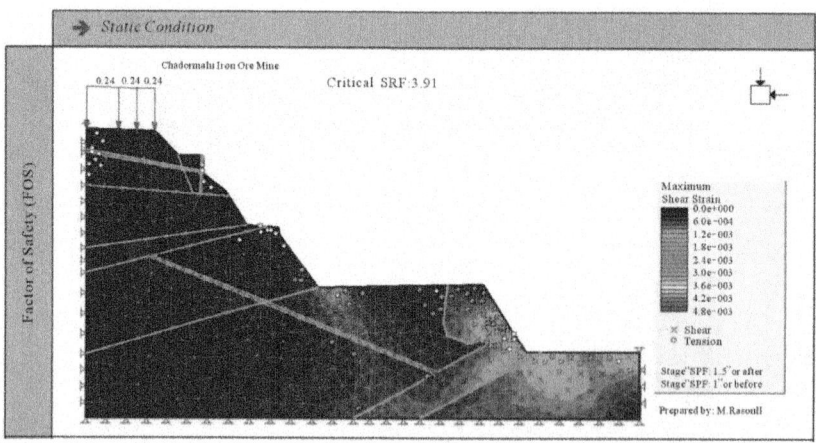

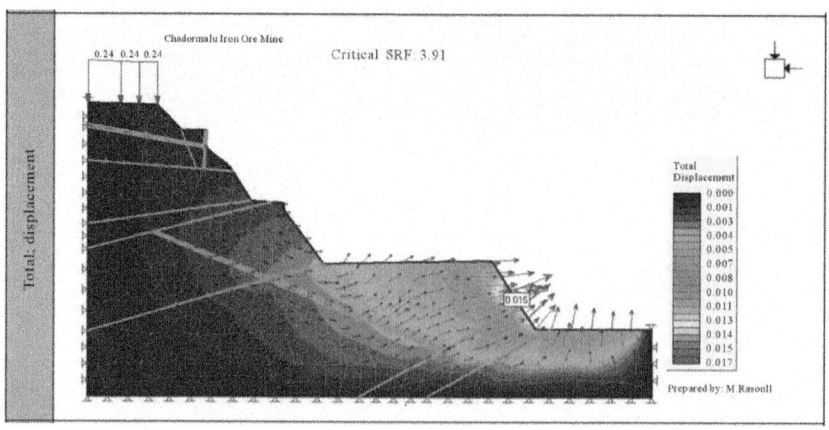

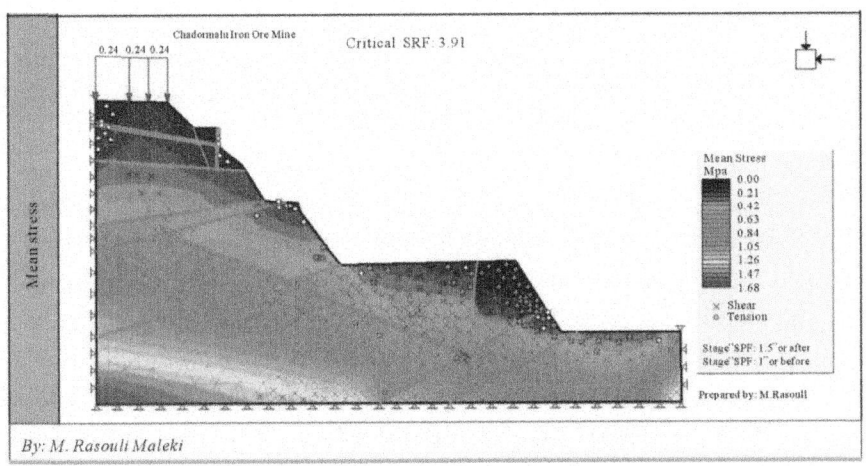

Figure 4: Output model of Phase2D software, factor of safety (FOS), total displacements and mean stress for dynamic state.

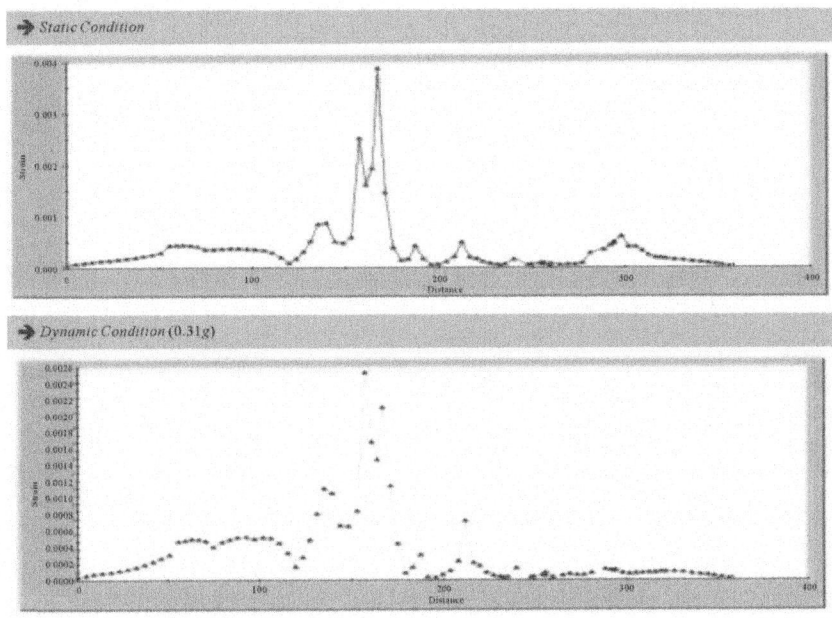

Figure 5: Shear-strain changes for static and dynamic states.

REFERENCES

1. Z. T. Bieniawski, "Engineering Rock Mass Classifications," Wiley, New York, 1989, pp. 5-249.

2.   M. Rasouli, "Study of the Engineering Geological Problems of the Havasan Dam, with Emphasis on Clay Filled Joints in the Right Abutment," International Journal of Rock Mechanics and Rock Engineering, 2011, pp. 1-16. doi:10.1007/s00603-011-0165-2

3.   M. Rasouli and M. Mahyar, "Assessment of Dominant Type of Failures in the Cutting of Transit Road Iran - Armenia Based on SMR Classification System," 4th National Conference on Rock Mechanics, Iran, 2011.

4.   M. Rasouli and M. Mahyar, "Assessment and comparison of occurring probability of rock failures based on empirical and kinematical methods," 4th National Conference on Rock Mechanics, Tehran, May 2011.

5.   M. Rasouli, "Assessment of Occurrence Probability for Planar & Wedge Failures under Dynamic & Static Conditions in Abutments of a Double Arch Concrete Dam (Case Study)," 4th International Conference on Geotechnical Engineering and Soil Mechanics, Tehran, 2-3 November 2010.

6.   Roc Science, Rock Mass Strength Analysis Using the Hoek-Brown failure criterion, 2005.

7.   Roc Science, Two-Dimensional Finite Element Slope Stability Analysis, 2005

CITATION

CHAPTER 1

Rekha R. Rao, Lisa A. Mondy, David R. Noble, Matthew M. Hopkins, Carlton F. Brooks and Thomas A. Baer (2012). 3D Numerical Modelling of Mould Filling of a Coat Hanger Distributer and Rectangular Cavity, Numerical Modelling, Dr. Peep Miidla (Ed.), ISBN: 978-953-51-0219-9,

CHAPTER 2

Bodhisattwa Chaudhuri, Fernando J. Muzzio and M. Silvina Tomassone (2011). Experimentally Validated Numerical Modeling of Heat Transfer in Granular Flow in Rotating Vessels, Heat Transfer - Mathematical Modelling, Numerical Methods and Information Technology, Prof. Aziz Belmiloudi (Ed.), ISBN: 978-953-307- 550-1.

CHAPTER 3

Leonid Bazyma, Vasyl Rashkovan and Vladimir Golovanevskiy (2012). Numerical Simulation of the Unsteady Shock Interaction of Blunt Body Flows, Numerical Modelling, Dr. Peep Miidla (Ed.), ISBN: 978-953-51-0219-9.

CHAPTER 4

Miroslaw Glowacki (2012). Inverse Analysis Applied to Mushy Steel Rheological Properties Testing Using Hybrid Numerical-Analytical Model, Numerical Modelling, Dr. Peep Miidla (Ed.), ISBN: 978-953-51-0219-9.

CHAPTER 5

Rubaiyet Iftekharul Haque, Christophe Loussert, Michelle Sergent, Patrick Benaben and Xavier Boddaert, Optimization of Capacitive Acoustic Resonant Sensor Using Numerical Simulation and Design of Experiment, doi: 10.3390/ s150408945.

CHAPTER 6

AmponDhamacharoen, (2016) Efficient Numerical Methods for Solving Differential Algebraic Equations. *Journal of Applied Mathematics and Physics*,**04**,39-47. doi: 10.4236/jamp.2016.41007

CHAPTER 7

Issakhov, A. (2013) Numerical Methods for Solving Turbulent Flows by Using Parallel Technologies. *Journal of Computer and Communications*, **1**, 1-5. doi: 10.4236/jcc.2013.11001.

CHAPTER 8

Ndiaye, C. , Fall, M. , Ndiaye, M. , Sangare, D. and Tall, A. (2014) A Review and Update of Analytical and Numerical Solutions of the Terzaghi One-Dimensional Consolidation Equation. *Open Journal of Civil Engineering*, 4, 274-284. doi: 10.4236/ojce.2014.43023.

CHAPTER 9

V. Niziev, F. Mirzade, V. Panchenko, M. Khomenko, R. Grishaev, S. Pityana and C. Rooyen, "Numerical Study to Represent Non-Isothermal Melt-Crystallization Kinetics at Laser-Powder Cladding," Modeling and Numerical Simulation of Material Science, Vol. 3 No. 2, 2013, pp. 61-69. doi: 10.4236/mnsms.2013.32008.

CHAPTER 10

G. Nut, I. Chiorean and P. Blaga, "Convergence and Error of Some Numerical Methods for Solving a Convection-Diffusion Problem," *Applied Mathematics*, Vol. 4 No. 5A, 2013, pp. 72-79. doi: 10.4236/am.2013.45A009.

CHAPTER 11

Abib Tall, Cheikh Mbow, Daouda Sangaré, Mapathé Ndiaye, Papa Sanou Faye, The Evolution of Pore Water Pressure in a Saturated Soil Layer between Two Draining Zones by Analytical and Numerical Methods, doi: 10.4236/ojce.2015.54039.

CHAPTER 12

M. Maleki, M. Mahyar and K. Meshkabadi, "Design of Overall Slope Angle and Analysis of Rock Slope Stability of Chadormalu Mine Using Empirical and Numerical Methods," *Engineering*, Vol. 3 No. 9, 2011, pp. 965-971. doi:10.4236/eng.2011.39119.

INDEX

A

Analyses of stability 267
Arvedi Steel Technology (AST) 105
Asymmetric energy 84, 86, 98, 101

C

Central composite design (CCD) 153
Control require 101

D

Design of experiments (DOE) 139, 141
Differential algebraic Equation (DAE) 167
Differential Algebraic Equation (DAE) 167
Direct numerical simulation (DNS) 180
Discrete Element Method (DEM) 33
Discretization method (FD) 240
Ductility recovery temperature (DRT) 107
Ductility Recovery Temperature (DRT) 107

F

Factor of Safety (FOS) 263
Filling process 1
Finite element method (FEM) 148, 240

F

Finite element methods (FEM) 139
Fraction dynamic 214

G

Geological Strength Index (GSI) 265

H

Hypersonic space vehicle 83

I

Infrastructure 243
Injection process 10, 13

L

Large eddy simulation (LES) 179, 180
Laser-powder cladding (LC) 203

M

Macroscopic scale 179
Material 31, 32, 33, 56, 58, 59, 61, 68, 70, 78, 80
Mathematical modeling 168
Matrix 228, 229, 230, 231, 232, 233, 234
Mechanical equilibrium 38, 54, 66
Melt pool dimensions 203
Melt pool dynamic 204
Method of resolution 247, 249

Molding process 5
Multigrid convergence 224
Multigrid method 223, 224, 227, 231,
        234, 235, 240

N

Nil strength temperature (NST) 107
Numerical analysis 192, 268
Numerical modeling 189, 211
Numerical simulation 250

P

Phenomenon of pressure 192, 198
Physical parameter 37
Polyethylene terephthalate (PET) 151

R

Response surface method (RSM 141

Rock Mass Rating (RMR) 265, 266

S

Sequential quadratic programming
        (SQP) 142
Slope Rock Mass Rating (SRMR) 265,
        266, 268
Strength Recovery Temperature (SRT)
        107

T

Technologie 105, 106, 107
Temperature control 2
Thermocouple 33
Thermodynamic 4
Total-variation-diminishing (TVD) 84
Transport process 179